# Advanced Studies
# Mobile Research Center Bremen

**Edited by**
O. Herzog,
C. Görg,
M. Lawo,
Bremen, Germany

Das Mobile Research Center Bremen (MRC) im Technologie-Zentrum Informatik und Informationstechnik (TZI) der Universität Bremen erforscht, entwickelt und erprobt in enger Zusammenarbeit mit der Wirtschaft mobile Informatik-, Informations- und Kommunikationstechnologien. Als Forschungs- und Transferinstitut des Landes Bremen vernetzt und koordiniert das MRC hochschulübergreifend eine Vielzahl von interdisziplinären Arbeitsgruppen, die sich mit der Entwicklung und Anwendung mobiler Lösungen beschäftigen. Die Reihe „Advanced Studies" präsentiert ausgewählte hervorragende Arbeitsergebnisse aus der Forschungstätigkeit der Mitglieder des MRC.

In close collaboration with the industry, the Mobile Research Center Bremen (MRC), a division of the Center for Computing and Communication Technologies (TZI) of the University of Bremen, investigates, develops and tests mobile computing, information and communication technologies. This research cluster of the state of Bremen links and coordinates interdisciplinary research teams from different universities and institutions, which are concerned with the development and application of mobile solutions. The series "Advanced Studies" presents a selection of outstanding results of MRC's research projects.

**Edited by**
Prof. Dr. Otthein Herzog
Prof. Dr. Carmelita Görg
Prof. Dr. Michael Lawo
Mobile Research Center, Bremen, Germany

Markus Becker

# Services in Wireless Sensor Networks

Modelling and Optimisation for
the Efficient Discovery of Services

Markus Becker
Bremen, Germany

Dissertation University of Bremen, 2013

mr**C**
Mobile Research Center

Gedruckt mit freundlicher Unterstützung des
MRC Mobile Research Center der Universität Bremen

Printed with friendly support of
MRC Mobile Research Center, Universität Bremen

ISBN 978-3-658-05401-4         ISBN 978-3-658-05402-1 (eBook)
DOI 10.1007/978-3-658-05402-1

The Deutsche Nationalbibliothek lists this publication in the Deutsche Nationalbibliografie; detailed bibliographic data are available in the Internet at http://dnb.d-nb.de.

Library of Congress Control Number: 2014934156

Springer Vieweg
© Springer Fachmedien Wiesbaden 2014
This work is subject to copyright. All rights are reserved by the Publisher, whether the whole or part of the material is concerned, specifically the rights of translation, reprinting, reuse of illustrations, recitation, broadcasting, reproduction on microfilms or in any other physical way, and transmission or information storage and retrieval, electronic adaptation, computer software, or by similar or dissimilar methodology now known or hereafter developed. Exempted from this legal reservation are brief excerpts in connection with reviews or scholarly analysis or material supplied specifically for the purpose of being entered and executed on a computer system, for exclusive use by the purchaser of the work. Duplication of this publication or parts thereof is permitted only under the provisions of the Copyright Law of the Publisher's location, in its current version, and permission for use must always be obtained from Springer. Permissions for use may be obtained through RightsLink at the Copyright Clearance Center. Violations are liable to prosecution under the respective Copyright Law.
The use of general descriptive names, registered names, trademarks, service marks, etc. in this publication does not imply, even in the absence of a specific statement, that such names are exempt from the relevant protective laws and regulations and therefore free for general use.
While the advice and information in this book are believed to be true and accurate at the date of publication, neither the authors nor the editors nor the publisher can accept any legal responsibility for any errors or omissions that may be made. The publisher makes no warranty, express or implied, with respect to the material contained herein.

Printed on acid-free paper

Springer Vieweg is a brand of Springer DE.
Springer DE is part of Springer Science+Business Media.
www.springer-vieweg.de

# Acknowledgement

I would like to thank my wife Elena Becker for the support during my time as research assistant, her entertainment and support. The grounding is important to electrical engineers!

Many thanks as well to my family Elisabeth and Ludwig Becker for the support and Lucia and Elias Ortmann.

I am thankful to my supervisors during my time at ComNets Prof. Carmelita Görg and Prof. Andreas Timm-Giel for the suggestions on the work and thesis, the fruitful discussions and so much more.

A big 'Thank you' to my colleagues at ComNets for the wonderful lunch time, all the big and small things: Bernd-Ludwig Wenning, Thushara Lanka Weerawardane, Koojana Kuladinithi, Asanga Udugama, Eugen Lamers, Li Xi, Chen Yi, Liang Zhao, Yasir Zaki, Amanpreet Singh, Mohammad Siddique, Andreas Könsgen, Varaporn Pangboonyanon, Chunlei An, Kaoru Yokoo, Umar Toseef, and Vo Que Son.

The work of the students, which I have supervised, has contributed and formed part of this thesis. Many thanks for your time and the fun we had: Matthias Gottwald, Liangliang Zhao, Liu Ye, Wei Wei, Uday Kumar Adusumilli, Yongzi Zeng, Yuntao Fu, Shaoping Yuan, Fahad Munawar Jan, Ban-Sok Shin, Zhiliang Chen, Jens Dede, Si Shuhao, Ronie Amin, Arturas Lukosius, Avanti Chitnis, Bo Jiehui, Fabian Monsees, Peter Trenkamp, Julian Schneider, Indika Sanjeewa, Samira Palipana, Felix Lüdeke.

Thanks also to my former students who joined ComNets: Gulshanara Sayyed (Singh) and Thomas Pötsch.

Everything that had to do with filling out travel applications or configuring the network and setting up hardware was kept off from me by the supporting staff of ComNets: Martina Kammann, Karl-Heinz Volk. Vielen Dank.

Many thanks to the colleagues of the CRC 637 (especially Martin Lorenz, Jan D. Gehrke, Christian Behrens and Reiner Jedermann), the Intelligent Container team (Adam Sklorz, Steffen Janßen, Alexander Dannies, Patrick Dittmer, Marius Veigt), COST 285 (notably Sebastien Rumley), ITG 5.2.4 (in particular Rastin Pries and Dr. Klaus Kohrt), the guest researchers at ComNets (Prof. Samir Das, Prof. Dirk Pesch, Prof. Ranjit Perera).

A huge thanks to my fellow students from Aachen for the big time we had back then and since then: Wolf-Martin Hoffmann, Jan Berger, Kevin Podratz, Christine Kirschfink, Bastian Memering and Philip Hafner and all their family members.

Another huge thanks to my friends: Oliver Thöne, Christoph Ammermann, Annika Schwartz, Philipp Linzner, and Sarah Potthast.

Thanks a lot to my supervisors during my student time who taught me a lot: Ingo Forkel, Ralf Pabst, Ole Klein, Marc Schinnenburg.

This research was partly supported by the German Research Foundation (DFG) as part of the Collaborative Research Centre 637 "Autonomous Cooperating Logistic Processes". This work has been part of the subprojects B3 'Mobile Kommunikationsnetze und -modelle' and T4 'Überwachungstechnologien für den Lebensmitteltransport'.

This research was additionally supported as part of the project 'The Intelligent Container' which is supported by the Federal Ministry of Education and Research, Germany, under reference number 01IA10001.

Many thanks to the industry partners for realising Wireless Sensor Networks for logistics with us: Axel Möhrke (Dole), Nils Schulte, Martin Schanzmann, Dieter Honkomp (Cargobull Telematics), Frank Drünert, Kurt Kretzschmar (OHB Teledata), Matthias Neugebauer (EMIC), Henri Kretschmer, Stefan Ziegler, Thomas Henn, Torsten Hueter (Virtenio), Dirk Unsenos (ISIS-IC), Fabian Pursche (ProSyst), Mehmet Kus, Frank Bittner and Ulfert Nehmiz (Otaris).

Markus Becker

# Abstract

Wireless Sensor Networks have been an active research item for the last decade. These networks consist of electronic devices which feature sensors to sense physical values, low-power microcontrollers to process them, memory to store data and low-power wireless communication means to transmit the data. Frequently, the networks are of a multi-hop nature to ensure sensing and communication coverage for a certain area.

The application areas of Wireless Sensor Networks are manyfold and reach from wild-life and habitat monitoring to industrial process control as well as from smart-home control to medical e-health applications. The application area of Wireless Sensor Networks covered in this thesis is the domain of transport logistics surveillance.

Originally Wireless Sensor Networks have been static and single-purpose. In recent years these networks have been moving towards applications that need support for mobility and multiple purposes. For example in logistical applications, such as transport condition surveillance, supervising equipment will be moved into and out of containers together with the supervised goods. Any of the involved parties in logistics has — possibly differing — interests in the data of the supervising equipment. This leads to heterogeneous applications and services on the supervising equipment in order to meet the differing interests.

These heterogeneous applications and services demand for a framework which distributes and discovers the various services, so that other pieces of equipment can use them. For an efficient and effective service discovery the algorithms for this distribution of services are of utmost importance, so that the framework can be used in application fields with diverse requirements.

Initially, this framework has been started by the author of this thesis with a simplistic algorithm, which proactively pushes services in a store-and-forward fashion among the nodes and removes the services from the local cache after a pre-defined duration. However, this approach is rather inefficient in static scenarios and ineffective in dynamic scenarios. An algorithm, which is efficient in static scenarios and also effective in dynamic scenarios has been proposed for other applications in the literature. This so called Trickle algorithm has been studied, extended, analytically modelled, simulated and employed in measurements in a Wireless Sensor Network testbed at the service layer in this thesis. The obtained results apply to the application of the Trickle algorithm at lower protocol layers as well. Given application delay requirements, the realisable distances and number of nodes for two network topologies have been derived from the 95 percentiles obtained by simulation.

The analytical model of the Trickle algorithm includes a model for the number of packets that are sent, which directly relates to the power that is spent. Additionally, a model for the time until a service is discovered for various network topologies is derived. It has been shown that the analytical models match the results from simulations for various network topologies (differing in number of nodes and in the distance between them as well as for different scenario layouts). The analytical model results can be obtained about $1/60^{th}$ of the time of the simulation and approximately $1/200^{th}$ of the time necessary for measurements.

It has been shown that service discovery frameworks can be efficiently and effectively employed in resource constrained Wireless Sensor Networks. The analytical models of the Trickle algorithm have been developed and gave insight into the behaviour of the algorithm. Previous published work focussed on studying the algorithm by means of simulations and measurements only. Furthermore it has been shown, that a non-adaptive algorithm (Regular Interval Pushing) can be parameterised to match either the delay characteristics or the overhead of the Trickle algorithm, but that it cannot match both metrics at the same time.

The service framework, employing the algorithm with its optimised parameters, is used in logistical feasibility studies in the research project 'The Intelligent Container' for the supervision of various transports with industrial research partners.

# Contents

| | |
|---|---|
| Contents | IX |
| List of Tables | XIII |
| List of Figures | XV |
| List of Abbreviations | XXI |
| List of Symbols | XXV |

**1 Introduction**     1
- 1.1 Logistics ................................ 1
- 1.2 Wireless Sensor Networks in Logistics ............ 1
- 1.3 Logistical Requirements of Wireless Sensor Networks .... 3
- 1.4 Services in Wireless Sensor Networks ............ 4
- 1.5 Objective ................................ 5
- 1.6 Application Scenarios ...................... 8
- 1.7 Contributions of this Thesis .................. 9
- 1.8 Overview ............................... 10

**2 State of the Art**     13
- 2.1 Wireless Sensor Networks .................... 13
  - 2.1.1 Organisations and Standardisation Bodies ...... 14
  - 2.1.2 Marketing Alliances .................... 24
  - 2.1.3 Research Community ................... 24
  - 2.1.4 Commercial Companies ................. 25
- 2.2 Service Discovery ......................... 26
  - 2.2.1 Service discovery protocols in Internet Protocol networks ... 26
  - 2.2.2 Service frameworks in WSN ............... 27
- 2.3 Trickle Algorithm ......................... 33
  - 2.3.1 Trickle Variables ..................... 33
  - 2.3.2 Trickle Constants .................... 34
  - 2.3.3 Trickle Rules ....................... 34

|  |  | 2.3.4 Trickle Applicability | 35 |

## 3 Service Distribution Algorithms — 37
- 3.1 Requirements of a Service Distribution Algorithm . . . . . . . . . 37
- 3.2 Flooding . . . . . . . . . . . . . . . . . . . . . . . . . . . . . . . . 37
- 3.3 Fixed Interval Pushing . . . . . . . . . . . . . . . . . . . . . . . . 38
- 3.4 Fixed Inteval Pushing with Vanish Support . . . . . . . . . . . . . 38
- 3.5 Trickle . . . . . . . . . . . . . . . . . . . . . . . . . . . . . . . . . 39
- 3.6 Trickle with Vanish Support . . . . . . . . . . . . . . . . . . . . . 39
- 3.7 Algorithm Comparison . . . . . . . . . . . . . . . . . . . . . . . . 40

## 4 Wireless Sensor Services Network Framework — 43
- 4.1 Service Layer Software Components . . . . . . . . . . . . . . . . . 45
- 4.2 Service Frame Format . . . . . . . . . . . . . . . . . . . . . . . . . 47
- 4.3 Service Forwarding in the Wireless Sensor Network . . . . . . . . 49
- 4.4 Internet Host Service Application . . . . . . . . . . . . . . . . . . 50

## 5 Evaluation Metrics and Scenarios — 51
- 5.1 Evaluation Metrics . . . . . . . . . . . . . . . . . . . . . . . . . . 51
  - 5.1.1 Number of Packets Sent . . . . . . . . . . . . . . . . . . . 51
  - 5.1.2 Energy Spent . . . . . . . . . . . . . . . . . . . . . . . . . 52
  - 5.1.3 Time to Consistency . . . . . . . . . . . . . . . . . . . . . 52
  - 5.1.4 Scalability with the Number of Nodes . . . . . . . . . . . 52
  - 5.1.5 Scalability with the Number of Services . . . . . . . . . . 52
- 5.2 Scenarios . . . . . . . . . . . . . . . . . . . . . . . . . . . . . . . 52
  - 5.2.1 Line Scenario . . . . . . . . . . . . . . . . . . . . . . . . . 53
    - 5.2.1.1 Only Direct Neighbours . . . . . . . . . . . . . 53
    - 5.2.1.2 Neighbours According to Propagation Model . . 53
  - 5.2.2 Grid Scenario . . . . . . . . . . . . . . . . . . . . . . . . . 53
  - 5.2.3 Random Scenario . . . . . . . . . . . . . . . . . . . . . . . 53
  - 5.2.4 Container Scenario . . . . . . . . . . . . . . . . . . . . . . 54

## 6 Simulation of Service Discovery — 59
- 6.1 Simulation Environment . . . . . . . . . . . . . . . . . . . . . . . 59
- 6.2 Simulation Evaluation . . . . . . . . . . . . . . . . . . . . . . . . 60
- 6.3 Simulation Results . . . . . . . . . . . . . . . . . . . . . . . . . . 60
  - 6.3.1 Statistical Significance . . . . . . . . . . . . . . . . . . . . 60
  - 6.3.2 Uniform Spatial Distribution of Sent Packets . . . . . . . 63
  - 6.3.3 Results for Grid and Random Scenarios . . . . . . . . . . 64

Contents XI

      6.3.4   Trickle Parameter Analysis . . . . . . . . . . . . . . . . . . 64
      6.3.5   95 Percentiles and Application Requirement Optimisation  66
      6.3.6   Trickle and Push Comparison . . . . . . . . . . . . . . . . 67
      6.3.7   Service Vanish Comparison . . . . . . . . . . . . . . . . . 69
      6.3.8   Routing Protocol Simulation Results . . . . . . . . . . . . 70

# 7 Analytical Modelling of Service Discovery   75
7.1  Models for the Service Distribution . . . . . . . . . . . . . . . . . 75
      7.1.1   Analytical Model for the Time to Consistency . . . . . . . 75
            7.1.1.1   Base Distributions $f_{h,c,a}(t)$ . . . . . . . . . . . 77
            7.1.1.2   Relative Frequency $p_{h,c,a}(t)$ . . . . . . . . . . 86
      7.1.2   Analytical Model for the Number of Packets Sent . . . . . 87

# 8 Measurements of Service Discovery in Wireless Sensor Networks   93
8.1  Measurement Setup . . . . . . . . . . . . . . . . . . . . . . . . . . 93
      8.1.1   Link Assessment Measurements . . . . . . . . . . . . . . 94
8.2  Measurement Results . . . . . . . . . . . . . . . . . . . . . . . . . 100

# 9 Evaluation   101
9.1  Comparison of Analytical Model and Simulation Results . . . . . 101
      9.1.1   Distribution Delay . . . . . . . . . . . . . . . . . . . . . . 102
      9.1.2   Mean Number of Sent Packets . . . . . . . . . . . . . . . 118
            9.1.2.1   Mean Number of Packets: Varying K, N=4, Line  118
            9.1.2.2   Mean Number of Packets: Varying K, N=64, Line 122
            9.1.2.3   Mean Number of Packets: Varying N, K=1, Line 124
            9.1.2.4   Mean Number of Packets: Varying N, K=3, Line 126
            9.1.2.5   Mean Number of Packets: Line/Grid, N=4, K=3 128
            9.1.2.6   Mean Number of Packets: Line/Grid, N=64, K=3 131
            9.1.2.7   Mean Number of Packets: Line/Grid, N=225, K=3 133
      9.1.3   Runtime behaviour . . . . . . . . . . . . . . . . . . . . . 135
9.2  Comparison of Measurement Results with Simulations and Analytical Model . . . . . . . . . . . . . . . . . . . . . . . . . . . . . . 135
      9.2.1   Delay . . . . . . . . . . . . . . . . . . . . . . . . . . . . 136
      9.2.2   Mean Number of Sent Packets . . . . . . . . . . . . . . . 140

# 10 Conclusions and Outlook   141
10.1  Conclusions . . . . . . . . . . . . . . . . . . . . . . . . . . . . . . 141
10.2  Outlook . . . . . . . . . . . . . . . . . . . . . . . . . . . . . . . . 143

# A Other contributions to communication networks research   145

| | | |
|---|---|---|
| **B** | **The Minimum of Several Random Variables** | **147** |
| **C** | **The Kaplan-Meier Estimator** | **149** |
| **D** | **Simulated PRR Topologies (Line-CPM)** | **151** |
| **E** | **Simulated PRR Topologies (Grid-CPM)** | **155** |
| **F** | **Radio Models** | **159** |
| | F.1 Signal Attenuation Model | 159 |
| | F.2 Packet Reception Ratio Model | 159 |
| **G** | **Approximating Step-Wise Linear Model** | **161** |
| | G.1 Approximating Linear Model for Line Scenarios | 161 |
| | G.2 Approximating Linear Model for Grid Scenarios | 161 |
| **Bibliography** | | **163** |

# List of Tables

| | | |
|---|---|---|
| 2.1 | WSN Hardware Platforms | 30 |
| 2.2 | Service Discovery Protocol Comparison (Extended from [Öst+06]) | 31 |
| 2.3 | Service Frameworks for Wireless Sensor Networks (WSNs) | 32 |
| 3.1 | Comparison of Algorithm Properties | 41 |
| 4.1 | Multicast Network Prefixes [Source: [Ded12]] | 49 |
| 4.2 | Multicast Interface Identifier [Source: [Ded12]] | 49 |
| 6.1 | Effect of Trickle Parameters (100 node Grid-CPM, 100 m Inter-Node Distance) [Source: [Ded12]] | 65 |
| 6.2 | Simulated 95th Percentiles of the Delay for the Line Scenarios in [s] | 72 |
| 6.3 | Simulated 95th Percentiles of the Delay for the Grid Scenarios in [s] | 73 |
| 6.4 | Comparison of Trickle and Push Algorithm [Source: [Ded12]] | 74 |
| 6.5 | Trickle Algorithm Parameter Values for RPL | 74 |
| 7.1 | PRR Matrix for a N=9, d=30m CPM Line Scenario | 89 |
| 7.2 | TX Outcome for a N=9, d=30m CPM Line Scenario | 90 |
| 7.3 | Hopcount probabilities for a N=9, d=30m CPM line scenario | 91 |
| 8.1 | Power Levels of the CC2420 Radio Chip [Source: [PP08]] | 95 |
| 9.1 | Trickle Algorithm Parameter Value | 101 |
| 9.2 | Runtime Comparison of the Applied Investigation Methods | 135 |
| 9.3 | Equivalence of Testbed Power Level and Inter-Node Distance in Simulation and Analytical Model | 139 |
| 9.4 | Comparison of the Mean Number of Sent Packets | 140 |
| C.1 | Confidence Level and Approximated Width of the Confidence Interval | 149 |

# List of Figures

| | | |
|---|---|---|
| 1.1 | Total World Delivered Energy Consumption by End-use Sector (Data from [US 10]) | 2 |
| 1.2 | Wireless Sensor Network for Logistics | 2 |
| 1.3 | Services in Logistical Applications | 6 |
| 1.4 | Forwarding of WSN services into the Internet Protocol (IP) network | 7 |
| 1.5 | Forwarding of IP services into the WSN | 7 |
| 2.1 | IEEE 802.15.4 frame format (From: [IEE03]) | 15 |
| 2.2 | Constellation diagram of IEEE 802.15.4 | 16 |
| 2.3 | Received IEEE 802.15.4 I/Q Digital Analog Converter (DAC) samples | 16 |
| 2.4 | 6LoWPAN Routing Options: Route Over | 19 |
| 2.5 | 6LoWPAN Routing Options: Mesh Under | 19 |
| 2.6 | Four 6LoWPAN Dispatch Headers | 29 |
| 2.7 | Trickle Period | 34 |
| 2.8 | Several Trickle Periods | 35 |
| 4.1 | Discovering Internet Services in the WSN | 44 |
| 4.2 | Discovering WSN Services in the Internet | 45 |
| 4.3 | Services Framework Protocol Stack | 46 |
| 4.4 | Services Framework Interfaces | 46 |
| 4.5 | Software Component Diagram: Service User | 46 |
| 4.6 | Software Component Diagram: Service Layer | 47 |
| 4.7 | Frame format (Short) | 48 |
| 4.8 | Frame format (Long) | 48 |
| 4.9 | Avahi Integration | 50 |
| 5.1 | Line-Direct Scenario with 9 Nodes | 54 |
| 5.2 | Line Scenario with 9 Nodes | 55 |
| 5.3 | Grid Scenario with 9 Nodes | 56 |
| 5.4 | Exemplary Random Scenario with 9 Nodes | 57 |
| 5.5 | Container Scenario | 58 |

| | | |
|---|---|---|
| 6.1 | Simulation Results for 9 Node Line-CPM Scenario with 95% Confidence Intervals . . . . . . . . . . . . . . . . . . . . . . . . . . . | 61 |
| 6.2 | Simulation Results for 25 Node Line-CPM Scenario with 95% Confidence Intervals . . . . . . . . . . . . . . . . . . . . . . . . . | 62 |
| 6.3 | Number of Sent Packets in 100 Node Grid Scenario Inter-Node Distance: 100 m, K=3, Observation Duration: 400 s [Source: [Ded12]] . . . . . . . . . . . . . . . . . . . . . . . . . . . . . . . . | 63 |
| 6.4 | Network Consistency Delay Distribution (Simulated, 16 nodes) . . | 65 |
| 6.5 | 16 Node Random Topology of one Monte Carlo Run . . . . . . . | 66 |
| 6.6 | Spiderweb Comparison of the Algorithms [Data Source: [Ded12]] | 68 |
| 6.7 | Comparison of Consistency Delay of Trickle and Push Algorithm [Source: [Ded12]] . . . . . . . . . . . . . . . . . . . . . . . . . . | 70 |
| 6.8 | Distribution of Version Numbers in a 100 Node Grid Scenario [Source: [Ded12]] . . . . . . . . . . . . . . . . . . . . . . . . . . | 71 |
| 6.9 | Time to Discovery of Default Route in the Container Scenario . . | 74 |
| 7.1 | Consistency Delay Addition . . . . . . . . . . . . . . . . . . . . | 76 |
| 7.2 | Node Consistency Delay Distribution . . . . . . . . . . . . . . . | 76 |
| 7.3 | Probability Density and Cumulative Distribution Function of the Network Consistency Delay Distribution . . . . . . . . . . . . . . | 79 |
| 7.4 | Probability Density Function for $h = 1, a = 0, c$ varied . . . . . . | 80 |
| 7.5 | Probability Density Function for $c = 0, a = 0, h$ varied . . . . . . | 81 |
| 7.6 | Probability Density Function for $c = 0, h$ and $a$ varied . . . . . . | 82 |
| 7.7 | Probability Density Function for $c = 0, h$ and $a$ varied . . . . . . | 83 |
| 7.8 | Probability Density Function for $h = 2, a = 0, c$ varied . . . . . . | 84 |
| 7.9 | Probability Density Function for $h = 3, a = 0, c$ varied . . . . . . | 85 |
| 7.10 | Packet Reception Ratio in a 2 node scenario with CPM propagation model . . . . . . . . . . . . . . . . . . . . . . . . . . . . . . . | 86 |
| 7.11 | 4 Node Scenario and Packet Reception Ratios between the Nodes | 87 |
| 8.1 | Wireless Sensor Node TelosB . . . . . . . . . . . . . . . . . . . . | 93 |
| 8.2 | Ceiling Network . . . . . . . . . . . . . . . . . . . . . . . . . . . | 94 |
| 8.3 | Measured Link-layer Packet Reception Ratios at Transmission Power Setting 0 . . . . . . . . . . . . . . . . . . . . . . . . . . . | 96 |
| 8.4 | Measured Link-layer Packet Reception Ratios at Transmission Power Setting 1 . . . . . . . . . . . . . . . . . . . . . . . . . . . | 97 |
| 8.5 | Measured link-layer Packet Reception Ratios at Transmission Power Setting 2 . . . . . . . . . . . . . . . . . . . . . . . . . . . | 98 |

List of Figures     XVII

| | | |
|---|---|---|
| 8.6 | Measured link-layer Packet Reception Ratios at Transmission Power Setting 3 | 99 |
| 9.1 | Network Consistency Delay Distribution (Line-CPM Scenario, 4 Nodes, K=3) | 103 |
| 9.2 | Network Consistency Delay Distribution (Line-CPM Scenario, 4 Nodes, K=1) | 104 |
| 9.3 | Network Consistency Delay Distribution (Line-CPM Scenario, 9 Nodes, K=3) | 105 |
| 9.4 | Network Consistency Delay Distribution (Line-CPM Scenario, 9 Nodes, K=1) | 106 |
| 9.5 | Network Consistency Delay Distribution (Line-CPM Scenario, 9 Nodes, K=9) | 107 |
| 9.6 | Network Consistency Delay Distribution (Line-CPM Scenario, 9 Nodes) | 108 |
| 9.7 | Network Consistency Delay Distribution (Line-CPM Scenario, 225 Nodes) | 109 |
| 9.8 | Network Consistency Delay Distribution (Line-CPM Scenario, 16 Nodes, K=3) | 110 |
| 9.9 | Network Consistency Delay Distribution (Line-CPM Scenario, 25 Nodes, K=3) | 111 |
| 9.10 | Network Consistency Delay Distribution (Grid-CPM Scenario, 4 Nodes, K=3) | 112 |
| 9.11 | Network Consistency Delay Distribution (Grid-CPM Scenario, 9 Nodes, K=3) | 113 |
| 9.12 | Network Consistency Delay Distribution (Grid-CPM Scenario, 225 Nodes) | 114 |
| 9.13 | Network Consistency Delay Distribution (Simulated, 9 Nodes) | 116 |
| 9.14 | Network Consistency Delay Distribution (Simulated, 225 Nodes) | 117 |
| 9.15 | Mean Number of Sent Packets in a 4 Node Line Scenario (K=1) (Observation Time: 400 s) | 119 |
| 9.16 | Mean Number of Sent Packets in a 4 Node Line Scenario (K=3) (Observation Time: 400 s) | 120 |
| 9.17 | Mean Number of Sent Packets in a 4 Node Line Scenario (K=9) (Observation Time: 400 s) | 121 |
| 9.18 | Mean Number of Sent Packets in a 64 Node Line Scenario (K=1) (Observation Time: 400 s) | 122 |
| 9.19 | Mean Number of Sent Packets in a 64 Node Line Scenario (K=3) (Observation Time: 400 s) | 123 |

9.20 Mean Number of Sent Packets in a 64 Node Line Scenario (K=9) (Observation Time: 400 s) .................... 123
9.21 Mean Number of Sent Packets in a 4 Node Line Scenario (K=1) (Observation Time: 400 s) .................... 124
9.22 Mean Number of Sent Packets in a 64 Node Line Scenario (K=1) (Observation Time: 400 s) .................... 125
9.23 Mean Number of Sent Packets in a 225 Node Line Scenario (K=1) (Observation Time: 400 s) .................... 125
9.24 Mean Number of Sent Packets in a 4 Node Line Scenario (K=3) (Observation Time: 400 s) .................... 126
9.25 Mean Number of Sent Packets in a 64 Node Line Scenario (K=3) (Observation Time: 400 s) .................... 127
9.26 Mean Number of Sent Packets in a 225 Node Line Scenario (K=9) (Observation Time: 400 s) .................... 127
9.27 Mean Number of Sent Packets in a 4 Node Line Scenario (K=3) (Observation Time: 400 s) .................... 129
9.28 Mean Number of Sent Packets in a 4 Node Grid Scenario (K=3) (Observation Time: 400 s) .................... 130
9.29 Mean Number of Sent Packets in a 64 Node Line Scenario (K=3) (Observation Time: 400 s) .................... 132
9.30 Mean Number of Sent Packets in a 64 Node Grid Scenario (K=3) (Observation Time: 400 s) .................... 132
9.31 Mean Number of Sent Packets in a 225 Node Line Scenario (K=3) (Observation Time: 400 s) .................... 134
9.32 Mean Number of Sent Packets in a 225 Node Grid Scenario (K=3) (Observation Time: 400 s) .................... 134
9.33 Delay, Power Setting 15 ..................... 136
9.34 Delay, Power Setting 3 ..................... 137
9.35 Delay, Power Setting 2 ..................... 138
9.36 Delay, Power Setting 1 ..................... 139

A.1 Internet Engineering Task Force (IETF) Protocol Stack for TinyOS 146

D.1 PRR Matrix (Distance 10m) ................... 151
D.2 PRR Matrix (Distance 50m) ................... 152
D.3 PRR Matrix (Distance 110m) .................. 153
D.4 PRR Matrix (Distance 145m) .................. 154

E.1 PRR Matrix (Distance 10m) ................... 155

## List of Figures

E.2 PRR Matrix (Distance 50m) . . . . . . . . . . . . . . . . . . 156
E.3 PRR Matrix (Distance 110m) . . . . . . . . . . . . . . . . . . 157
E.4 PRR Matrix (Distance 145m) . . . . . . . . . . . . . . . . . . 158

F.1 Packet Reception Ratio (PRR)-Signal to Noise Ratio (SNR) Model 160

# List of Abbreviations

| | | |
|---|---|---|
| **6LoWPAN** | Internet Protocol Version 6 over Low power Wireless Personal Area Networks | 8 |
| **API** | Application Programming Interface | 27 |
| **ASK** | Amplitude Shift Keying | 17 |
| **BGP** | Border Gateway Protocol | 146 |
| **blip** | Berkeley low-power IP implementation | 8 |
| **BMBF** | Bundesministerium für Bildung und Forschung | 9 |
| **BPSK** | Binary Phase Shift Keying | 15 |
| **BSL** | Bootstrap Loader | 25 |
| **BTU** | British Thermal Unit | 1 |
| **CBPS** | Content-Based Publish Subscribe | 27 |
| **CCA** | Clear Channel Assessment | 14 |
| **ccdf** | Complementary Cumulative Distribution Function | XXVII |
| **cdf** | Cumulative Distribution Function | 52 |
| **CI** | Continuous Integration | 146 |
| **CPU** | Central Processing Unit | 17 |
| **CPM** | Closest-Fit Pattern Matching | 25 |
| **CoAP** | Constrained Application Protocol | 20 |
| **CoRE** | Constrained RESTful Environments | 20 |
| **COST** | European Cooperation in Science and Technology | 145 |
| **CRC** | Collaborative Research Center | 1 |
| **CSMA** | Carrier Sense Multiple Access | 17 |
| **CSMA/CA** | Carrier Sense Multiple Access / Collision Avoidance | 14 |
| **DA** | Directory Agent | 31 |
| **DAC** | Digital Analog Converter | XV |
| **DFG** | Deutsche Forschungsgemeinschaft | 145 |

| | | |
|---|---|---|
| **DHCP** | Dynamic Host Configuration Protocol | 49 |
| **DIO** | Destination Oriented Directed Acyclic Graph Information Object | 20 |
| **DNS** | Domain Name Service | 17 |
| **DSSS** | Direct Sequence Spread Spectrum | 15 |
| **DTLS** | Datagram Transport Layer Security | 24 |
| **ED** | Energy Detection | 14 |
| **ETSI** | European Telecommunications Standards Institute | 146 |
| **EU** | European Union | 146 |
| **FEFO** | First Expire First Out | 3 |
| **FIFO** | First In First Out | 3 |
| **HTML** | Hypertext Markup Language | 18 |
| **HTTP** | Hypertext Transfer Protocol | 18 |
| **I** | in-phase | 15 |
| **ICT** | Information and Communication Technology | 1 |
| **IEEE** | Institute of Electrical and Electronics Engineers, Inc. | 8 |
| **IETF** | Internet Engineering Task Force | XVIII |
| **IP** | Internet Protocol | XV |
| **IPSO** | Internet Protocol for Smart Objects | 24 |
| **IPv4** | Internet Protocol Version 4 | 17 |
| **IPv6** | Internet Protocol Version 6 | 13 |
| **ISO** | International Organization for Standardization | 43 |
| **LLN** | Low power and Lossy networks | 20 |
| **LPL** | Low Power Listening | 17 |
| **LR-WPAN** | Low-Rate Wireless Personal Area Network | 14 |
| **LQI** | Link Quality Indicator | 14 |
| **MAC** | Medium Access Control Sublayer | 5 |
| **mDNSv6** | Multicast Domain Name System for IPv6 | 49 |
| **mDNS-SD** | Multicast Domain Name System - Service Discovery | 6 |
| **MEMS** | Micro Electro-Mechanical Systems | 13 |
| **MRHOF** | Minimum Rank Objective Function with Hysteresis | 70 |

| | | |
|---|---|---|
| MTU | Maximum Transmission Unit | 18 |
| nesC | Network Embedded Systems C | 25 |
| NoE | Network of Excellence | 146 |
| OF0 | Objective Function 0 | 70 |
| OMA | Open Mobile Alliance | 146 |
| OS | Operating System | 24 |
| OSI | Open Systems Interconnection | 43 |
| O-QPSK | Offset Quadrature Phase-Shift Keying | 15 |
| PHY | Physical Layer | 15 |
| pdf | Probability Density Function | 52 |
| PRNG | Pseudo-Random Number Generator | 59 |
| PRR | Packet Reception Ratio | XIX |
| PSSS | Parallel Sequence Spread Spectrum | 17 |
| Q | quadrature | 15 |
| RAM | Random Access Memory | 135 |
| RegExp | Regular Expression | 60 |
| REST | Representational State Transfer | 20 |
| RFC | Request for Comment | 8 |
| RFID | Radio Frequency Identification | 1 |
| ROLL | Routing Over Low power and Lossy networks | 20 |
| RPL | IPv6 Routing Protocol for Low power and Lossy Networks | 20 |
| RTC | Real Time Clock | 5 |
| RV | Random Variable | 78 |
| SA | Service Agent | 31 |
| SD | Service Discovery | 9 |
| SDR | Software Defined Radio | 15 |
| SDRP | Service-Driven Routing Protocol | 28 |
| SLP | Service Location Protocol | 26 |
| SNR | Signal to Noise Ratio | XIX |
| SOAP | Simple Object Access Protocol | 18 |

| | | |
|---|---|---|
| **SOSANET** | Service-Oriented Sensor-Actuator Network | 27 |
| **SRE** | Service Request Extension | 27 |
| **SRpE** | Service Reply Extension | 27 |
| **SSLP** | Simple Service Location Protocol | 26 |
| **T** | Type | 21 |
| **TCP** | Transmission Control Protocol | 13 |
| **TEP** | TinyOS Enhancement Proposal | 25 |
| **TinyOS** | Tiny Operating System | 8 |
| **TKL** | Token Length | 21 |
| **TOSSDR** | TinyOS Software Defined Radio | 146 |
| **TOSSIM** | TinyOS Simulator | 25 |
| **TTL** | Time to Live | 38 |
| **TXT** | DNS TXT Record | 48 |
| **UDP** | User Datagram Protocol | 17 |
| **UMTS** | Universal Mobile Telecommunication System | 145 |
| **UPnP** | Universal Plug and Play | 8 |
| **URI** | Uniform Resource Identifier | 20 |
| **USB** | Universal Serial Bus | 6 |
| **USRP** | Universal Software Radio Peripheral | 146 |
| **Ver** | Version | 21 |
| **WLAN** | Wireless Local Area Network | 145 |
| **WSN** | Wireless Sensor Network | XIII |
| **WSNs** | Wireless Sensor Networks | 10 |
| **XML** | Extensible Markup Language | 26 |

# List of Symbols

| | | |
|---|---|---|
| $\alpha$ | Confidence level | 150 |
| $\beta_1$ | PRR model parameter | 159 |
| $\beta_2$ | PRR model parameter | 159 |
| $\gamma$ | Propagation coefficient | 159 |
| $\hat{S}(t)$ | Estimate of $S(t)$ | 150 |
| $\mathcal{L}$ | Laplace Transform | 78 |
| $\mathcal{L}^{-1}$ | Inverse Laplace Transform | 78 |
| $\mathcal{O}(N)$ | Linear Scalability with the Number of Nodes | 52 |
| $\mathcal{O}(S)$ | Linear Scalability with the Number of Services | 52 |
| $\mathrm{var}\{\hat{S}(t)\}$ | Variance of $\hat{S}(t)$ | 150 |
| $\overline{E^*}$ | Mean Energy Spent (Per Node) | 52 |
| $\overline{M^*}$ | Mean Number of Packets (Per Node) | 52 |
| $\overline{M}$ | Mean Number of Packets (Complete Network) | 52 |
| $\overline{M}$ | Mean Number of Packets in $Tm$ | 88 |
| $\overline{N^n}$ | Mean Number of Neighbours | 88 |
| $\sigma$ | Standard deviation | 150 |
| $\tau$ | Trickle Communication Interval Length | 33 |
| $\tau_H$ | Highest Trickle Communication Interval Length | 34 |
| $\tau_L$ | Lowest Trickle Communication Interval Length | 34 |
| $\Theta(\cdot)$ | Heaviside Step Function | 78 |
| $a$ | Number of 1-hop Ancestors Closer to the Source | 76 |

| | | |
|---|---|---|
| $C$ | Trickle Communication Counter | 33 |
| $c$ | Trickle Cycle | 76 |
| $C'$ | Maximum Number of Trickle Cycles to take into Account | 76 |
| $C_V$ | Trickle Vanish Version Counter | 39 |
| $d$ | Distance (Signal Attenuation Model) | 159 |
| $d$ | Inter-Node Distance | 59 |
| $d_i$ | Number of Deaths at Time $t_i$ | 149 |
| $E$ | Energy Spent (Complete Network) | 52 |
| $E_i^*$ | Energy Spent (Per Node) | 52 |
| $f_{h,c,a}(t)$ | Base Distributions for the Analytical Model | 76 |
| $h$ | Hops | 76 |
| $I$ | RFC 6206 Symbol for $\tau$ | 33 |
| $Imax$ | RFC 6206 Symbol for $\tau_H$ | 34 |
| $Imin$ | RFC 6206 Symbol for $\tau_L$ | 34 |
| $K$ | Trickle Redundancy Constant | 34 |
| $M$ | Number of Packets (Complete Network) | 52 |
| $M_i^*$ | Number of Packets (Per Node) | 52 |
| $N$ | Noise (PRR Model) | 159 |
| $N$ | Number of Nodes | 52 |
| $N_i^n$ | Number of Neighbours of Node $i$ | 88 |
| $n_i$ | Number of Survivors just prior to $t_i$ | 149 |
| $P(T \leq t)$ | Cumulative Distribution Function | 52 |
| $p(t)$ | Probability Density Function | 52 |
| $p_{h,c,a}(t)$ | Relative Frequency of the Base Distributions for the Analytical Model | 76 |

# List of Symbols

| | | |
|---|---|---|
| $phi(\cdot)$ | Normalised normal distribution | 150 |
| $PL_0$ | Pathloss at 1 m distance | 159 |
| $PRR$ | Packet Reception Ratio | 159 |
| $S(t)$ | A Complementary Cumulative Distribution Function (ccdf) | 150 |
| $S$ | Number of Services | 52 |
| $S_{RX}$ | Received Signal Strength | 159 |
| $S_{TX}$ | Transmission Power | 159 |
| $T$ | Trickle Timer Value | 33 |
| $t$ | Time | 52 |
| $t_i$ | $i^{th}$ Time | 149 |
| $T_m$ | Observation Period | 88 |
| $t_{100\%}$ | Time at which 100% of the Nodes are Informed | 161 |
| $t_{95\%}$ | Time at which 95% of Services are Detected | 74 |
| $t_{max}$ | Time at which the Maximum Number of Services are Detected | 74 |
| $t_{push}$ | Push Interval Length | 38 |
| $t_{rm}$ | Time of Removal of a Service | 69 |
| $V$ | Trickle Vanish Version Number | 39 |
| $V_{missed}$ | Trickle Vanish Maximum Missed Version Numbers | 39 |
| $w$ | Multiple of the standard deviation $\sigma$ | 150 |
| $Y$ | Number of Trickle Cycles in $Tm$ | 88 |
| $n_{\neg RX}$ | Not-Receiving Nodes | 90 |
| $n_{RX}$ | Receiving Nodes | 90 |
| $n_{TX}$ | Transmitting Node | 90 |
| $P$ | Probability for a Certain Transmission Outcome | 90 |

# 1 Introduction

## 1.1 Logistics

Logistics is a multi-player business which has changed significantly in the last decade. The changes are driven by several factors, e.g. by smaller batch sizes (because of customisation and individual orders) or by technological changes (Radio Frequency Identifications (RFIDs), WSNs). Information technology becomes an integral part of logistics and helps in lowering costs, while additionally allowing supervision of the transport conditions of the goods. Changing to a more decentralised control of logistic processes and application of information technology in logistics is the topic of the inter-disciplinary Collaborative Research Center (CRC) 637 'Autonomous Cooperating Logistics Processes – A Paradigm Shift and its Limitations' [Col].

The energy spent world-wide summed up to approximately 370 quadrillion British Thermal Units (BTUs) (see Glossary) in 2008 and is expected to increase to more than 520 quadrillion BTUs in 2035 as projected by the U.S. Department of Energy in [US 10] shown in figure 1.1. Transportation logistics represents a major part in the transportional sector (the other major part is private transport) and production logistics is an important factor to the industrial sector as well. The transportation and industrial sectors together are expected to rise from 290 to 430 quadrillion BTUs in the same time frame and exceed the amount of energy spent by the residential and commercial sectors. Modern Information and Communication Technology (ICT) can help in conserving energy, improving efficiency and reducing costs of logistical processes by reducing unnecessary transports, improved information on transportation conditions and adaptation of energy expenditure to the transportation conditions.

## 1.2 Wireless Sensor Networks in Logistics

The application of WSNs in logistics, e.g. transport of food and perishable goods, for surveillance of the transport conditions is economically and ecologically beneficial. Fig. 1.2 shows one application scenario, where WSN nodes are attached to goods (mostly food because of their perishable nature). The goods are loaded from a warehouse to a freight vehicle, in which the attached nodes need to self-organise

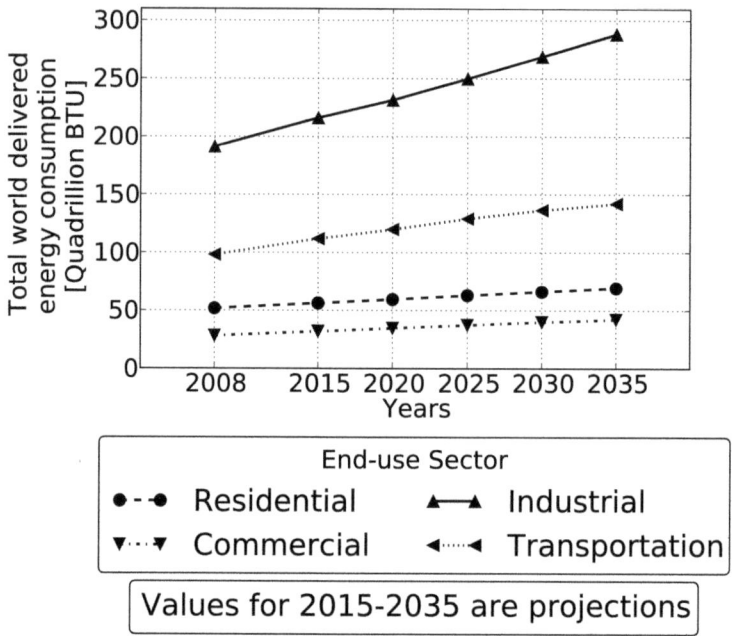

**Figure 1.1:** Total World Delivered Energy Consumption by End-use Sector (Data from [US 10])

and form a network of nodes, which can deliver information of the goods' state to the outside world by using a gateway (e.g. a telematic unit).

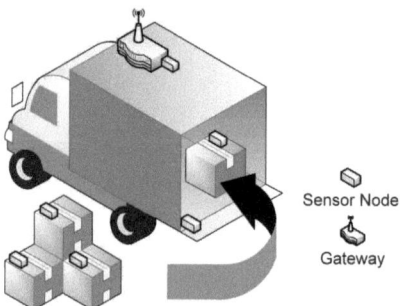

**Figure 1.2:** Wireless Sensor Network for Logistics

A WSN as information source is an enabler for a new type of logistics, e.g. dynamic First Expire First Out (FEFO). Contrary to the currently employed First In First Out (FIFO) or static First Expire First Out (FEFO) strategy which uses delivery dates or static best before dates, dynamic FEFO takes the dynamically changing best before dates into account by using information acquired during storage and transport [Jed09]. As the temperature logging of goods in transport is legally required in the European Union [Amt04] and standards for temperature loggers are existing [Eur99], the usage of WSNs and their dynamic data enable more sophisticated logistic applications.

The improvements by a better perishable goods transport logistic, e.g. food logistics, are among others the reduction of spoilt goods, an improved quality of the goods and a better visibility of risks along the transport chain of the goods. The additionally acquired information can be fed into production planning systems for further data mining.

Logistics benefits clearly from WSNs. However, the requirements of logistics for applicable WSNs are challenging.

## 1.3 Logistical Requirements of Wireless Sensor Networks

One of the challenges of general-purpose WSNs in logistics includes self-configuration of sensor nodes especially in the case of mobility (which is typically present in logistics). When a node enters a new WSN, e.g. as in figure 1.2, several initialisation steps are necessary. First of all, the node needs to configure the correct radio channel in order to be able to communicate with neighbouring nodes and the gateway. When the radio channel is not known ex ante, the node could either detect this by channel activity sensing on all possible channels or it could be configured out-of-band (e.g. by a handheld). Additionally the node has to figure out, where to send its data (to a database at the gateway, to a database in the network behind the gateway or to a database in the Internet).

WSNs for logistics are very likely not tailor-made single-purpose WSNs, but general-purpose WSNs with several tailor-made services. Logistics involves many parties (e.g. senders, shippers, carriers, receivers). The WSN nodes are thus of different ownerships (e.g. the gateway node belongs to the container owner, the freight supervision node is owned by the sender or the receiver), are possibly of different hardware platforms or have different supervision algorithms (depending on the good to be supervised). A WSN in logistics would therefore be made up out of nodes which are greatly varying. Tailor-made WSNs would not be applicable, but standardised protocols which allow for dynamic reconfiguration are necessary.

In addition to standardised physical layers, medium access control, and networking protocols, a mechanism for solving the typical dynamic application layer problems in logistical applications is needed. One mechanism to solve dynamic application layer problems is a service discovery protocol. The protocol is distributing available services in the network. Nodes could then discover services at their current location and could reconfigure themselves to integrate in the present network.

The supervision of the real world by using WSNs is progressing rapidly. The virtual world of the Internet therefore is enriched with data from the real world. The integration of this data, however, is not yet well supported, as the networks for the supervision and the networks for data transmission are considerably different.

## 1.4 Services in Wireless Sensor Networks

Research in WSNs has made significant progress over the last years. This research has been centered on hardware platforms and the lower layers of the protocol stack, especially the physical, link and network layers, allowing for very small energy-efficient networked nodes (1-2 cm range and less [Rei02]). The integration of the nodes and their services by higher layers of the protocol stack and the dynamic remote access, which is of importance to many application domains, has been neglected by research. A service and service discovery approach for sensor networks will allow the autonomous integration of services in the WSNs as well as across network borders. Autonomous and zero-configuration operation is, e.g., necessary for application domains such as cargo surveillance in transport logistics, as it is not feasible that the users manually configure large scale networks, mostly due to time constraints.

The concept of the service is in wide use in IP networks, e.g. in the Service Oriented Architecture [PL03] and Service Discovery Protocols [BR00]. The concept has been relatively neglected in WSNs so far.

Services in WSNs will help to improve the usage of WSNs by non-researchers, e.g. users from the logistics industry. Dynamic WSN services, because of their loose coupling, are integrated more easily into information systems of the industry than statically pre-assigned services. Since WSNs in logistics are always in flux by the nature of logistics, loose coupling offers better extensibility support than tight coupling, where all information systems need to be updated at the same time.

Based on the evaluation of current service protocols in IP networks, service discovery protocols, which discover services across the WSN and Internet boundary, are designed and implemented. The algorithms for the discovery are analytically evaluated with regard to the overhead created by the protocol, the scalability with the number of nodes and services, and the performance of the service discovery.

The usability improvements by the service protocol can significantly enhance the application of WSNs in logistics and telematics as well as other application domains, which are in need of autonomous and self-configuring sensor networks, which can be easily integrated with backend services in the Internet and corporate networks.

The design of the architecture and protocols needs to take into account that WSNs are usually massively distributed systems. The network nodes need to act autonomously and should not rely on single nodes (e.g. service directories on only one single node); rather the massive distribution should be exploited in the architecture and the protocols.

Different service announcement strategies are studied together with the adaptation of the strategies to the amount of nodes present in the WSN and the individual service needs. The key characteristics to be studied are: presence of remote services, up-to-dateness of remote services, and overhead introduced by the service framework.

## 1.5 Objective

The main objective of this thesis is to develop energy-efficient self-organised service discovery algorithms. One of the main example scenarios will focus on logistical applications. Exemplary services which are relevant for logistical applications are shown in figure 1.3.

Exemplary services of Wireless Sensor Networks are:

- Measurement services (e.g. humidity, temperature, gas concentration, door opening)

- Identification services (e.g. Medium Access Control Sublayer (MAC) addresses, authorisation tokens, certificates)

- Gatewaying services (e.g. routes to corporate networks)

- Database services (availability of permanent storage databases)

- Data processing services (e.g. conversion between different data formats such as JSON, XML)

- Time service (provides date and time to nodes without Real Time Clocks (RTCs))

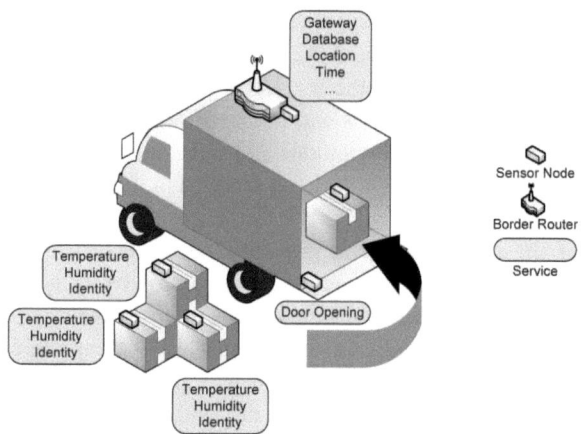

**Figure 1.3: Services in Logistical Applications**

The discovery of such services has to be solved in a generic way as pointed out earlier. This thesis aims at solving this using service discovery protocols. The algorithms, which are part of the protocols, are the main scientific research object of this thesis. One example of an algorithm, which can be used for service discovery, is the so called *Trickle* algorithm. Improvements of this algorithm and adaptations for the specific case of service discovery will be performed. Other algorithms will be evaluated and used for comparison.

The performance of these algorithms will be analysed by means of analytical modelling, simulation and experiments. The algorithms have to scale with the number of wireless sensor nodes, the number of services and the topology of the WSN.

Furthermore, the solution needs to enable service discovery between WSNs and Internet Protocol Networks, so that nodes in a local-area IP network can discover services in the WSN (cf. figure 1.4) and vice versa as shown in figure 1.5.

In step 1 of figure 1.4 the service is announced by the service functionality of the *ServiceApplication*. The service is received by the *PppRouter* application on the gateway device and forwarded over the Universal Serial Bus (USB) port (step 2). The proxy application *mdns-proxy.py* which runs on the host computer gets the service announcement and announces the service using Multicast Domain Name System - Service Discovery (mDNS-SD) into the Local Area Network (step 3), where it can be detected by other hosts (step 4). After detection, the services can be used.

## 1.5 Objective

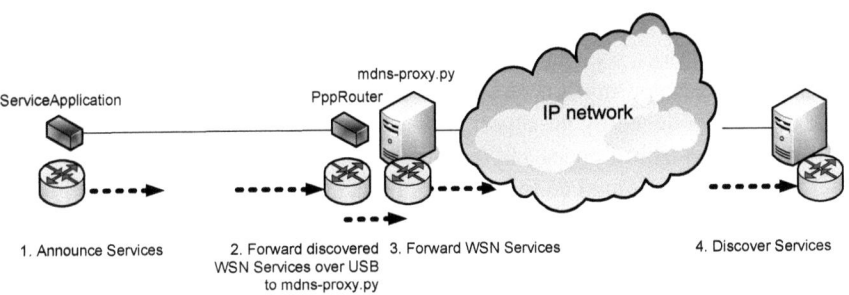

**Figure 1.4: Forwarding of WSN services into the IP network**

Figure 1.5 shows the opposite way: transmitting services from IP networks to Wireless Sensor Networks, is depicted. An IP host announces a service (4) and the gateway host's mDNS-SD proxy mdns-proxy.py recognises it. The discovered service is forwarded over the USB port to the WSN node (3), which broadcast the IP service into the WSN network (2). Eventually, one or more sensor nodes receive the service announcement (1) and are able to use it.

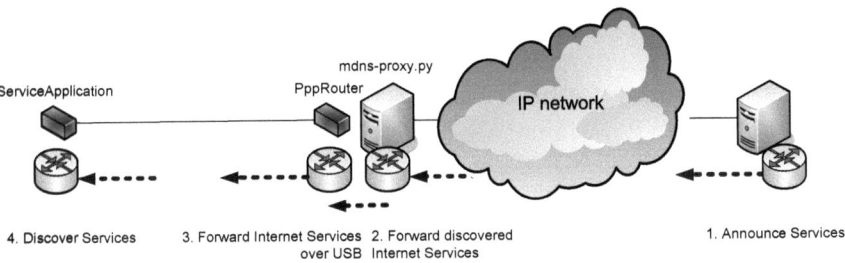

**Figure 1.5: Forwarding of IP services into the WSN**

In logistical applications special sensors are for example gas detection sensors and door opening sensors, which are required for food environment conditions monitoring [BB65] and container or warehouse security violation detection [DS08]. Simple WSN nodes with common sensors have to deliver the data to an IP network. The destination of the data is likely to change before, during and after transport, so that the need for the discovery of the delivery location is obvious. The shown use cases are just two examples of many use cases in logistics and even more in other application domains (e.g. smart grids). A generic solution for across-network service discovery thus has high applicability for the shown and mentioned use cases.

As many networks, especially in logistics, are going to be battery powered, the service discovery algorithms need to be energy-efficient. The overhead in energy-expenditure introduced by the algorithms needs to be limited and needs to scale well with the number of nodes and services. The overhead consists of service discovery packets transmitted additionally to the data packets, thus using additional energy. The consumed energy and the amount of packets will be studied as some of the main performance indicators. Additionally the time to discover newly added nodes and their respective services will be investigated. The methods employed are measurements, simulations and analytical modeling.

For the dissemination of the algorithms into the application domains, it is important that the solution, in which the algorithms are embedded, is based on already present standards and that the solution itself is going to be standardised. Industrial users are more likely to adopt standardised solutions. Especially in an application domain such as logistics standardised solutions are necessary because of the many partners involved in logistics processes and the physical exchange of goods and sensor nodes.

The implementation of efficient service discovery algorithms in this thesis are based on:

- standards such as Institute of Electrical and Electronics Engineers, Inc. (IEEE) 802.15.4 [IEE03; IEE06], IETF Request for Comment (RFC) 4944 Internet Protocol Version 6 over Low power Wireless Personal Area Networks (6LoWPAN) [Mon+07], Multicast Domain Name System - Service Discovery (mDNS-SD) as specified in IETF RFC 6762 [CK13b] and 6763 [CK13a], Universal Plug and Play (UPnP) [ISO07]

- open source software such as Tiny Operating System (TinyOS) and the 6LoWPAN implementation Berkeley low-power IP implementation (blip)

- the adaptive algorithm Trickle [Lev+04] which was first published in [Lev+04] and later standardised in RFC 6206 [Lev+11].

## 1.6 Application Scenarios

The work presented in this thesis is guided by an application scenario validating the approach taken and the solutions found. The application scenario under consideration is transport and warehouse logistics. It was selected as a realistic scenario for future deployment of WSNs with an economical and ecological impact and is a challenging scenario with mobility of nodes and networks, dense and sparse sensor node populations, leading to challenging network topologies with a high degree of dynamicity in certain periods of time. The same requirements can also be found in other application fields, like advanced metering and healthcare. The scenario is used to illustrate the feasibility of service discovery in WSNs.

The application scenarios are detailed with their major players, roles and applications. Optional and mandatory requirements derived from the application scenarios are listed and described. The logistical application scenario is the focus scenario to be researched.

In the research project 'The Intelligent Container: Linked Intelligent Objects in Logistics (2010-2013)' funded by the Bundesministerium für Bildung und Forschung (BMBF) this scenario has been successfully implemented by the Communication Networks group of the University of Bremen together with

- the industrial application partners: Dole, Cold Chain Group and its daughter company Rungis Express as well as Kühn Transporte Lagerung,

- the industrial technology partners: Cargobull Telematics, OHB Teledata, European Microsoft Innovation Center, ISIS IC, Otaris Interactive Services, Virtenio, aicas, CHS Spezialcontainer Shelter and Engineering, Elbau Elektronik Bauelemente, ProSyst, Seeburger Business Integration, Texas Instruments,

- and the research partners: ATB Leibniz Institute for Agricultural Engineering Potsdam-Bornim, the working group Cold Chain Managment of the Institute of Animal Science at the University Bonn as well as following institutes of the University Bremen: Institute for Microsensors, -actuators und -systems, Institute for Theoretical Elektrical Engineering and Microelectronics, and Bremen Institute for Production und Logistics,.

In the project several successful field tests of the Wireless Sensor Network (WSN) and other technology components have been conducted in the logistical supply chains of the application partners. With Dole, a WSN and a telematic by OHB Teledata had supervised the transport conditions inside a cargo container fully loaded with bananas from Latin America to Europe across the Atlantic Ocean. At Kühn and Rungis Express, a freight vehicle transporting meat was supervised with a WSN and a telematic device by Cargobull Telematics.

## 1.7 Contributions of this Thesis

In this thesis the feasiblity of Service Discovery (SD) in WSNs has been shown alongside with the exchange of services between WSNs and IP networks. The solution has been demonstrated multiple times to the scientific community.

The underlying algorithms for the distribution of the services have been studied in-depth. The main contribution is the creation of analytical models for the so-called Trickle algorithm. The evaluated criteria are the time for the distribution of the service as well as the number of sent packets.

The models for the algorithm can be applied to code distribution and the distribution of routing layer information as well as for service distribution. Since the Trickle algorithm is used in the IETF RFC 6550 'RPL: IPv6 Routing Protocol for Low-Power and Lossy Networks', the findings of this thesis can be used to trade the time until routes are found off against the number of overhead packets necessary to distribute the routes.

The Trickle algorithm is compared to a non-density-adaptive interval pushing algorithm. A flooding algorithm has not been considered because of the well-known drawback of creating broadcast storms. An extension to the Trickle algorithm has been realised to support the detection of when nodes and their services have vanished.

For the evaluation of the algorithm and the service discovery several scenarios have been set up to study the behaviour and gain insight into adaquate parameterisation of the algorithm. The parameters with significant influence are deduced and their impact is shown.

The statistical confidence of the simulation results has been shown. The good fit of the analytical model with the simulation results and the measurement results is additionally supporting the validity of the results.

During the creation of the thesis a variety of tools have been created, e.g. to allow the simulation of the Internet Protocol based Wireless Sensor Network protocols, and to evaluate the simulation results.

## 1.8 Overview

The thesis is structured in the following way: In chapter 2 the state of the art for Wireless Sensor Networks (WSNs), Service Discovery as well as the important algorithm Trickle are presented. Chapter 3 lists and describes requirements for an algorithm for the distribution of services into a network and introduces the algorithms. The framework, in which the algorithms are used, is presented in chapter 4. The metrics, which are used to evaluate the algorithms, are presented together with the evaluation scenarios in chapter 5. The evaluation of the algorithms by means of simulation is introduced in chapter 6, by detailing the simulation environment, the evaluation procedure and the simulation results. Analytical models, which have

## 1.8 Overview

been developed, are introduced and derived in chapter 7. The measurement setup and the measurement results of the framework and the algorithm are presented in chapter 8. Chapter 9 contains the evaluation of the results of the analytical modelling, the simulation and the measurements as a comparison. The conclusions are drawn and the outlook is presented in chapter 10.

# 2 State of the Art

## 2.1 Wireless Sensor Networks

Wireless Sensor Networks are networks of wirelessly communicating nodes which have the ability to measure physical variables. The individual nodes are usually embedded devices consisting of a microcontroller, memory, sensors, a wireless network interface, a programming interface and batteries. The development in the field of Micro Electro-Mechanical Systems (MEMS) has led to miniaturisation of the nodes down to the size of less than 1 cm$^3$ [Har+07].

The vision that is pursued by the WSN community is to connect the physical world with the virtual world [Est+02]. Information, i.e. measurements from real-world phenomena, is made available in a virtual world in order to benefit from this information. This vision can be applied in many application fields, e.g. Nature Monitoring, Smart Home/Office, Logistics, Agriculture, Traffic Management, and Healthcare.

In order to be economically and/or technically feasible WSN nodes are restricted with respect to the available resources, e.g. energy, processing power, communication bandwidth. The nodes therefore typically need to be custom designed in terms of hardware and software to the desired application. In this thesis the custom design is reduced for the software of the nodes and based on standard protocols which can be configured to meet the user's requirements.

The integration of WSNs and IP networks as an enabler to the integration of data into business processes is a precondition to a wide acceptance of Wireless Sensor Networks. WSNs in itself are useful – a tighter integration with IP networks would improve the applicability to more application areas, e.g. logistics.

While Wireless Sensor Networks (WSNs) are a recent and active research area, IP networks have been around for several decades, are widely accepted and used, and still under further improvement.

The interconnection between WSNs and IP networks has typically been performed by transport level gateways (e.g. TinyOS Serial Forwarder). The data from the WSN is extracted over the serial or USB port and can be fetched by other applications on a Transmission Control Protocol (TCP) port. Recently network layer gatewaying between WSN and the Internet has been made feasible, so that IP packets can be fed into WSNs. The IETF 6LoWPAN working group has made

Internet Protocol Version 6 (IPv6) in WSNs possible by RFC 4944 [Mon+07]. The advantages of the availability of an IP adaptation layer for WSNs are among others [SB10]:

- the interoperability with other IP networks (many control systems support IP as an option)
- the presence of well established tools for network management (e.g. ping, telnet, traceroute)
- trusted security solutions (firewalls, access control).

In the following the organisations, communities and standardisation activities which are relevant to this thesis and Wireless Sensor Networks (WSNs) will be presented.

### 2.1.1 Organisations and Standardisation Bodies

Professional organisations which are working in the field of WSNs are coming from standardisation bodies such as the IEEE, the IETF or from marketing alliances such as the ZigBee Alliance and the Internet Protocol for Smart Objects Alliance or the research community as well as indivual companies.

**Institute of Electrical and Electronics Engineers**

The IEEE has standardised the protocol 802.15.4 in 2003 [IEE03] and enhanced it in 2006 [IEE06], under the umbrella of the 802.15 working group. The standards describe the Physical and Medium Access Control Layer for Low-Rate Wireless Personal Area Networks (LR-WPANs). This standard is implemented in a variety of low-power chipsets, e.g. the Chipcon CC2420 and the Atmel AT86RF230.

The physical layer of 802.15.4 takes care of the listed functionalities:

- Energy Detection (ED) on the chosen channel
- Clear Channel Assessment (CCA) for implementing Carrier Sense Multiple Access / Collision Avoidance (CSMA/CA)
- Link Quality Indicator (LQI) on received packets
- Powering and unpowering the radio transceiver, channel switching and data transmission/reception

## 2.1 Wireless Sensor Networks

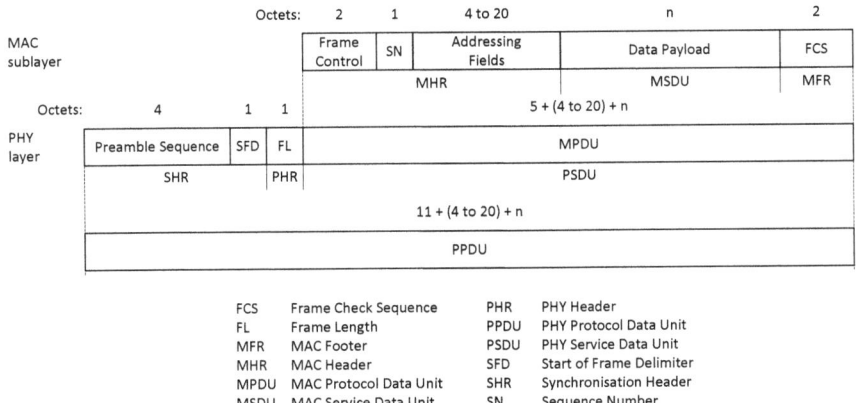

Figure 2.1: IEEE 802.15.4 frame format (From: [IEE03])

The MAC and Physical Layer (PHY) frame format of IEEE 802.15.4 is shown in figure 2.1.

The 802.15.4 standard defines the usage of two frequency bands at 868/915 MHz (continent dependent) and 2400 MHz (internationally). The lower band, although having a better radio propagation, is not easily usable for international logistics as the higher layers would have to detect in which location it is currently operated to meet the regulatory requirements (cf. Annex F of [IEE03] and [IEE06]). Additionally, only few radio chips are available for the lower bands. The 2400 MHz band, however, is internationally available and regulations on this band are more homogeneous, so that it can be seen as a good fit for internationally deployed WSNs in logistics.

The coding and modulation on those bands has originally been specified in the 802.15.4-2003 standard to be Direct Sequence Spread Spectrum (DSSS) with Binary Phase Shift Keying (BPSK) with 20/40 kbit/s datarate for the 868/915 MHz and DSSS with Offset Quadrature Phase-Shift Keying (O-QPSK) with 250 kbit/s on the 2.4GHz band. A constellation diagram of O-QPSK created with a Software Defined Radio (SDR) is shown in figure 2.2, while figure 2.3 shows the in-phase (I) and quadrature (Q) symbols for the start of an IEEE 802.15.4 packet transmission. The I symbols are modulated on the half-sine pulse shape carrier, while the Q symbols are delayed by half the chip duration and modulated on a cosine carrier.

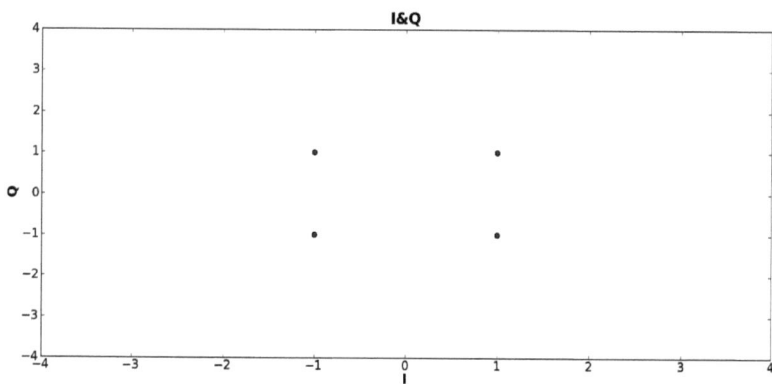

**Figure 2.2:** Constellation diagram of IEEE 802.15.4

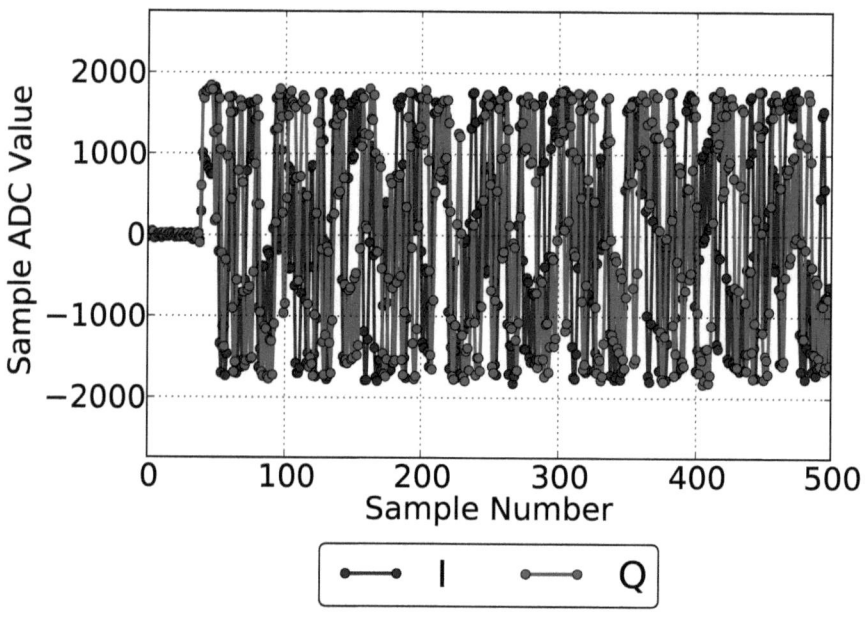

**Figure 2.3:** Received IEEE 802.15.4 I/Q DAC samples

The 2006 standard added DSSS with O-QPSK and Parallel Sequence Spread Spectrum (PSSS) with BPSK and Amplitude Shift Keying (ASK) modulation, thereby introducing higher data rates for the lower band [IEE06].

The MAC layer which is described by the standards mentioned above is usually not in use in programs based on TinyOS, though implementations exist [Hau09] for specific hardware [KH08]. In TinyOS typically a Carrier Sense Multiple Access (CSMA) with random backoffs and possibly a Low Power Listening (LPL) duty-cycle algorithm [ML08] is used.

**Internet Engineering Task Force**

The Internet Engineering Task Force (IETF) is standardising networking solutions on top of link-layer protocols in the form of RFCs. Due to the specific nature (i.e. low throughput, lossy links and small packet size) of the 802.15.4 layers, the adaptation of the widely used IP protocol needs to be specified. As solutions involving IEEE 802.15.4 based devices are likely to be deployed in the hundreds or thousands of nodes per installation, the addressing space available for the nodes needs to be of adequate size. The addressing space of Internet Protocol Version 4 (IPv4) [Pos80] is limited to less than $2^{32}$ and is currently almost depleted [Hus09] due to the inefficient usage of the addressing space. Thus IPv6 [DH95] with its address space of $2^{128}$ addresses was chosen to be the networking protocol to be adapted to the underlying link layer protocols, as detailed in the following paragraph.

The IETF Working Group 6LoWPAN has standardised the transmission of IPv6 packets in IEEE 802.15.4 Wireless Sensor Networks [Int; TZI]. The working group has created the problem statement document RFC 4919 [KMS07] and the specification of the frame format in the document RFC 4944 [Mon+07] and RFC 6282 [HT11].

The challenges of implementing IPv6 on WSN nodes are to meet the critical embedded wireless requirements of long lifetime under the constraint of limited energy, high reliability and adaptability as well as manageability of many devices with highly constrained resources (i.e. Central Processing Unit (CPU) power, memory size).

The advantages that are prevalent when using IP in WSNs include interoperability with existing IP network links (e.g. Ethernet and 802.11), security support by established mechanisms (e.g. Firewalls and Authentication Schemes), existing name-address translation (i.e. Domain Name Service (DNS)), inter-working with mobility protocols (e.g. MobileIP), end-to-end transport protocols (TCP and User Datagram Protocol (UDP) among others) and many application and management protocols and applications (Hypertext Transfer Protocol (HTTP), Hypertext Markup Language (HTML), Simple Object Access Protocol (SOAP), ping, traceroute, etc.) [SB10].

The most important challenges imposed by layer 2 (IEEE 802.15.4) are the small Maximum Transmission Unit (MTU) of maximum 127 bytes while IP requires support for 1280 bytes MTUs, from which the need for fragmentation support and header (including addresses) compression can be inferred. This is handled by the definition of an adaptation layer. RFC 4944 defines the packet format for transmission of IPv6 RFC 2460 [DH98] frames. Additionally the creation of IPv6 link-local addresses as well as statelessly autoconfigured addresses for IEEE 802.15.4 networks are defined.

A basic UDP/IPv6/802.15.4 header has a length of 58 bytes, of which UDP/IPv6 make up 48 bytes. However, those headers contain redundant information across the layers, especially the length information and addresses (IPv6 addresses can be derived from 802.15.4 addresses), which can be removed (reduction by 20.5 bytes). The compression of commonly used values (flow labels, address prefixes, multicast addresses, UDP ports and checksum) can be compressed (reduction by 24.5 bytes). Thus in the best case the UDP / IPv6 header can be compressed to 6 bytes, according to [HC08]. In more detail, when using link-local unicast the UDP/IPv6 header is 6 bytes, when using link-local multicast 7 bytes and when using global unicast the header has a length of 10 bytes [HC08].

**Routing** 6LoWPAN allows for the forwarding and routing processes to be implemented at the link layer (termed 'Mesh Under') and at the network layer (termed 'Route Over').

**Route Over** In Route Over forwarding and path computation is performed by layer 3, i.e. the IP layer above the 6LoWPAN adaptation layer. Every node in this scheme is an IP router. The architectural view of this scheme is depicted in figure 2.4. With this option the IP link spans only over one-hop radio transmission range.

**Mesh Under** In Mesh Under forwarding and path computation is performed by layer 2, so either in the 6LoWPAN adapation layer or in the data link layer below the adaptation layer (both options are possible). Every node in this scheme is an 802.15.4 switch. The architectural view of this scheme is depicted in figure 2.5. With this option the IP link spans over several radio transmission ranges. Mesh Under emulates transitivity across the link for the IP layer, which means that if

## 2.1 Wireless Sensor Networks

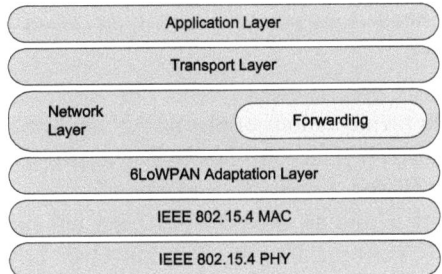

Figure 2.4: 6LoWPAN Routing Options: Route Over

node A and node B as well as node B and node C can communicate, then node A and node C can communicate as well.

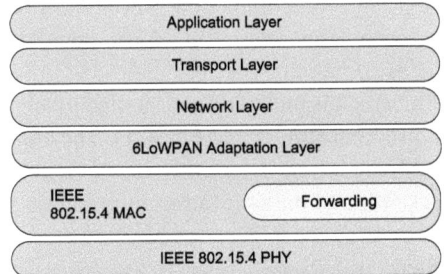

Figure 2.5: 6LoWPAN Routing Options: Mesh Under

**6LoWPAN header format** The 6LoWPAN frame format starts with a Dispatch byte. The various Dispatch options are shown in fig. 2.6.

If the Dispatch byte begins with '00', the packet is not a 6LoWPAN frame. If the first two bits are '01', a normal IPv6 addressing header is following. When '10' are in the first two bits, a mesh header is following. In the case of '11' in the first bits, a fragmentation header is following.

A normal IPv6 addressing header can have uncompressed or various compressed modes for the addressing and other header information.

The mesh header allows for the 'Mesh Under' mode of 6LoWPAN relying on meshing capabilities of the underlying protocols.

The fragment header is present when the service data unit and the 6LoWPAN header would not fit into one link-layer packet and indicate the fragment number.

If multiple of these headers are present in one frame, the order is mesh, fragment and then normal header.

**Implementations** Implementations of 6LoWPAN are available from ArchRock [Arc07], Sics (SicsLowpan) [Dur+08], Jacobs University [HS08], Berkeley University (blip) [Jia+09; HC08], SensiNode and others.

The specification of a routing protocol has been left out of the 6LoWPAN working group and is performed in the Routing Over Low power and Lossy networks (ROLL) working group. The working group ROLL started by assembling requirements documents for several application domains, i.e. for Home Automation [BBP10], Industrial Applications [Pis+09] and Urban Environments [Doh+09]. Additional requirements of 6LoWPAN on the routing protocol were stated in [Kim+12].

ROLL has specified a routing protocol for Low power and Lossy networks (LLN), termed IPv6 Routing Protocol for Low power and Lossy Networks (RPL), in the standardisation documents from RFC 6550 up to RFC 6554 [Win+12; Vas+12; Thu12; HV12; Hui+12].

The Trickle algorithm which is modelled and studied in this thesis has been standardised by the ROLL group in RFC 6206 [Lev+11]. The algorithm is used for the distribution of so called Destination Oriented Directed Acyclic Graph Information Object (DIO) messages, which contain routing information that is necessary for the discovery of routes towards the RPL root node, which is in most cases also the 6LoWPAN border router.

The Constrained RESTful Environments (CoRE) working group of the IETF is working on the standardisation of an application level protocol for constrained nodes and networks (such as WSNs) termed Constrained Application Protocol (CoAP), which is based on the Representational State Transfer (REST) architecture. The current status of the CoAP protocol is specified in [She+13]. The protocol is based on UDP, implements its own retransmission scheme and has a more compact representation than HTTP, while also defining proxying between HTTP and CoAP.

Using link-formats under a well-known Uniform Resource Identifier (URI) (i.e. /.well-known/core) allows for the discovery of other available resources. Resource discovery and service discovery (as described in this thesis) complement each other. E.g. service discovery can be used to discover the nodes while resource discovery is then used to discover the resources on the nodes. The service discovery described in this thesis additionally can perform in a fully decentralised fashion. Whereas the resource discovery follows the client-server model more strictly.

In the following the main data formats and identifiers of CoAP are summarised

## 2.1 Wireless Sensor Networks

based on [She+13]. The extended version of this summary can also be found in [Bec13].

The CoAP Message Format is shown in listing 1, where Ver denotes Version, T denotes Type, and TKL stands for Token Length. Code either includes the CoAP method code in requests or the CoAP response code in a response. The Message ID field gives a unique identifier for a CoAP message to be matched with the response. A token is used for CoAP response matching when the response is separate response with a different Message ID. Options might be included in the CoAP messages, the end of the options is marked by a payload marker of eight '1' bits (0xFF). The payload is appended at the end.

```
 0                   1                   2                   3
 0 1 2 3 4 5 6 7 8 9 0 1 2 3 4 5 6 7 8 9 0 1 2 3 4 5 6 7 8 9 0 1
+-+-+-+-+-+-+-+-+-+-+-+-+-+-+-+-+-+-+-+-+-+-+-+-+-+-+-+-+-+-+-+-+
|Ver| T |  TKL  |      Code     |          Message ID           |
+-+-+-+-+-+-+-+-+-+-+-+-+-+-+-+-+-+-+-+-+-+-+-+-+-+-+-+-+-+-+-+-+
|   Token (if any, TKL bytes) ...
+-+-+-+-+-+-+-+-+-+-+-+-+-+-+-+-+-+-+-+-+-+-+-+-+-+-+-+-+-+-+-+-+
|   Options (if any) ...
+-+-+-+-+-+-+-+-+-+-+-+-+-+-+-+-+-+-+-+-+-+-+-+-+-+-+-+-+-+-+-+-+
|1 1 1 1 1 1 1 1|    Payload (if any) ...
+-+-+-+-+-+-+-+-+-+-+-+-+-+-+-+-+-+-+-+-+-+-+-+-+-+-+-+-+-+-+-+-+
```

**Listing 1: CoAP Message Format (Source: [She+13])**

The method types in the CoAP message formats' field T can be the values listed in listing 2.

```
+------+-----------------+
| Type | Name            |
+------+-----------------+
|   0  | CONfirmable     |
|   1  | NON-confirmable |
|   2  | ACKnowledgement |
|   3  | ReSeT           |
+------+-----------------+
```

**Listing 2: CoAP Method Types (Source: [She+13])**

The CoAP method codes for requests are similar to the HTTP verbs and are listed in listing 3. The method codes are included in the field Code of the CoAP message format.

```
+------+--------+
| Code | Name   |
+------+--------+
|    1 | GET    |
|    2 | POST   |
|    3 | PUT    |
|    4 | DELETE |
+------+--------+
```

**Listing 3: CoAP Method Codes (Source: [She+13])**

In listing 4 the structure of CoAP codes for responses is shown, while in listing 5 the classes of response codes and in listing 6 the various codes and the equivalent HTTP code and meaning are listed. The response code can be found in the field Code of the CoAP message format.

```
 0
 0 1 2 3 4 5 6 7
+-+-+-+-+-+-+-+-+
|class|  detail |
+-+-+-+-+-+-+-+-+
```

**Listing 4: CoAP Response Code Structure (Source: [She+13])**

```
+-------+--------------+
|Class  |              |
+-------+--------------+
| 2.xx  | Success      |
| 4.xx  | Client Error |
| 5.xx  | Server Error |
+-------+--------------+
```

**Listing 5: CoAP Response Code Classes (Source: [She+13])**

## 2.1 Wireless Sensor Networks

```
+------+--------------------------------+
| Code | Description                    |
+------+--------------------------------+
|   65 | 2.01 Created                   |
|   66 | 2.02 Deleted                   |
|   67 | 2.03 Valid                     |
|   68 | 2.04 Changed                   |
|   69 | 2.05 Content                   |
|  128 | 4.00 Bad Request               |
|  129 | 4.01 Unauthorized              |
|  130 | 4.02 Bad Option                |
|  131 | 4.03 Forbidden                 |
|  132 | 4.04 Not Found                 |
|  133 | 4.05 Method Not Allowed        |
|  134 | 4.06 Not Acceptable            |
|  140 | 4.12 Precondition Failed       |
|  141 | 4.13 Request Entity Too Large  |
|  143 | 4.15 Unsupported Content-Format|
|  160 | 5.00 Internal Server Error     |
|  161 | 5.01 Not Implemented           |
|  162 | 5.02 Bad Gateway               |
|  163 | 5.03 Service Unavailable       |
|  164 | 5.04 Gateway Timeout           |
|  165 | 5.05 Proxying Not Supported    |
+------+--------------------------------+
```

**Listing 6: CoAP Response Code Classes (Source: [She+13])**

The options that can be used in CoAP messages are shown in listing 7. The abbreviations in the table are: C: Critical, U: Unsafe, N: No-Cache-Key, and R: Repeatable.

```
+-----+---+---+---+---+----------------+--------+--------+---------------+
| No. | C | U | N | R | Name           | Format | Length | Default       |
+-----+---+---+---+---+----------------+--------+--------+---------------+
|   1 | x |   |   | x | If-Match       | opaque | 0-8    | (none)        |
|   3 | x | x | - |   | Uri-Host       | string | 1-255  | (see          |
|     |   |   |   |   |                |        |        | core-coap-13) |
|   4 |   |   |   | x | ETag           | opaque | 1-8    | (none)        |
|   5 | x |   |   |   | If-None-Match  | empty  | 0      | (none)        |
|   7 | x | x | - |   | Uri-Port       | uint   | 0-2    | (see          |
|     |   |   |   |   |                |        |        | core-coap-13) |
|   8 |   |   |   | x | Location-Path  | string | 0-255  | (none)        |
|  11 | x | x | - | x | Uri-Path       | string | 0-255  | (none)        |
|  12 |   |   |   |   | Content-Format | uint   | 0-2    | (none)        |
|  14 |   | x | - |   | Max-Age        | uint   | 0-4    | 60            |
|  15 | x | x | - | x | Uri-Query      | string | 0-255  | (none)        |
|  16 |   |   |   | x | Accept         | uint   | 0-2    | (none)        |
|  20 |   |   |   | x | Location-Query | string | 0-255  | (none)        |
|  35 | x | x | - |   | Proxy-Uri      | string | 1-1034 | (none)        |
|  39 | x | x | - |   | Proxy-Scheme   | string | 1-255  | (none)        |
+-----+---+---+---+---+----------------+--------+--------+---------------+
```

**Listing 7: CoAP Options (Source: [She+13])**

The format and type of the payload in CoAP requests or responses is specified by the Content-Format/Media-Type as given in listing 8.

```
+----------------------------+-----+
| Media type                 | Id. |
+----------------------------+-----+
| text/plain;charset=utf-8   |   0 |
| application/link-format    |  40 |
| application/xml            |  41 |
| application/octet-stream   |  42 |
| application/exi            |  47 |
| application/json           |  50 |
+----------------------------+-----+
```

**Listing 8: CoAP Content-Formats (Source: [She+13])**

CoAP URIs are similarly structured as HTTP URIs and are shown in listing 9. The coaps:// scheme denotes secure communication using Datagram Transport Layer Security (DTLS).

```
coap-URI  = "coap:"  "//" host [ ":" port ] path-abempty [ "?" query ]
coaps-URI = "coaps:" "//" host [ ":" port ] path-abempty [ "?" query ]
```

**Listing 9: CoAP URI schemes (Source: [She+13])**

### 2.1.2 Marketing Alliances

**ZigBee Alliance**

The ZigBee Alliance [Ziga] standardised vertical protocol solutions to enable products. Philips, ember, Schneider Electric, and Texas Instruments are among the promoters and participants of the ZigBee Alliance [Zigb].

Recently the ZigBee alliance has defined an extension of their software stack onto 6LoWPAN in the ZigBee Smart Energy 2.0 specification [Zig10].

**IP Smart Objects Alliance**

Among the members of the Internet Protocol for Smart Objects (IPSO) Alliance are ArchRock, Atmel, Bosch, Ericsson, Intel, SAP, sensinode and Sun. The aim of the alliance is to promote the usage of IP in Smart Objects and create awareness of the applicability and availability of IP in embedded systems [IPS]. The IPSO Alliance is hosting interoperability events for their members.

### 2.1.3 Research Community

**TinyOS**

TinyOS [Tin07; Lev+05] is an open-source Operating System (OS) for WSNs, developed at the University of California, Berkeley, in collaboration with Intel Research. Since its beginnings in 1999 it has attracted a community, that has ported

TinyOS to currently 13 supported platforms (e.g. micaZ, TelosB). Its development is documented in so called TinyOS Enhancement Proposals (TEPs), which define the major core components and their software interfaces. TinyOS is an event based OS and its interfaces are split-phase, i.e. a method call usually does not immediately return a result, but the result is signalled back later. TinyOS uses the programming language Network Embedded Systems C (nesC) which enforces the event signalling. The applications are preprocessed into regular C code by nesC's compiler, which eliminates code fragments, which are not used. The OS and the application code are statically linked together and create the executable, which is installable by a so called Bootstrap Loader (BSL). TinyOS contains several sample and test applications as well as the TinyOS Simulator (TOSSIM).

TOSSIM [Lev+03] is a discrete event simulator for TinyOS applications. The TinyOS application is compiled to a shared library, which can be loaded by a Python simulation control script. One limitation of TOSSIM is the restriction to one application per simulation, which means that all nodes need to run the same application. By runtime checks of the node number, different behaviours of the simulated nodes can be achieved though. The hardware of the sensor nodes is replaced by software implementations. Closest-Fit Pattern Matching (CPM) [LCL07] can be used as a radio model. Other simplistic radio models can be used as well. The simulated application needs to be enhanced with debugging output statements, which when executed are written to files. The debugging output can be post-processed after the simulation execution.

Other OSs for WSNs are e.g. Contiki [DGV04] and SOS [Han+05].

### 2.1.4 Commercial Companies

Many platforms for WSNs have been designed, developed and produced by universities and companies. The most prominent ones are Crossbow/MEMSIC Mica, MicaZ, Mica2 and TelosB [Cro], Moteiv (now Sentilla) TMoteSky [Sena], and Sun Spots [Sun]. Other notable systems are Ambient Systems' SmartPoint, Router and Gateway [Amb], Sensinode's devices, NanoRouter and NanoGateway [Senb], ArchRock's systems [Arc] Zolertia's TelosB clone Z1, and Virtenio's Preon32. A comparison of a selection of WSN hardware platforms is listed in table 2.1.

MEMSIC TelosB devices have been used for the measurements performed for this thesis, because of the good platform and driver support in the TinyOS as well as the Contiki operating system. Additionally, the device has been found to have a well-designed and well-working radio subsystem [TBG11].

## 2.2 Service Discovery

Service discovery protocols allow an easy discovery of services in a network. Different methods are possible for the discovery, e.g. announcing services (Push), requesting services (Pull) or publish-subscribe mechanisms. Currently, service discovery protocols are implemented, e.g., in networked printers [Hew04], which announce their availability into the network (Push).

The common solutions for Internet Protocol (IP) based networks are extremely resource demanding, so that they are not feasible in this form for the resource constrained devices in WSNs. The devices are particularly constrained in terms of memory, computational power, communication bandwidth and energy. Thus new and adapted methods are necessary in this domain.

### 2.2.1 Service discovery protocols in Internet Protocol networks

Service Discovery protocols have been in wide use in IP networks. There is a wide variety of protocols, e.g., Universal Plug and Play (UPnP) [ISO07], Multicast Domain Name System - Service Discovery (mDNS-SD) [CK13b; CK13a], Service Location Protocol (SLP) [Gut+99], and Jini [Sun07]. These service protocols have been compared and evaluated in [BR00].

Table 2.2 (which has been extended from a table in [Öst+06]) compares the predominant service discovery protocols. Jini [Sun07], as a centrally oriented and Java-based solution, is not feasible for sensor networks. SLP [Gut+99] and its derivation Simple Service Location Protocol (SSLP) [Kim+07] are not regarded as candidates for the service discovery in sensor networks, as only the Pull technique is supported and the current implementations in the market are not in widespread use. Full UPnP [ISO07] is not possible on sensor network devices as stated in [Son+05], and needs to be adapted (which has been done in the same publication). However, the authors of [Son+05] created a single point of failure in their gateway point: as they implicitly assume to have only one gateway. The mDNS-SD Bonjour protocol supports decentralised operation [CK13b] and there is a possibility to have several gateways between the WSN and IP networks. The content of the service messages is encoded in the well-known DNS format and can be processed by sensor nodes (contrary to Extensible Markup Language (XML)). Both techniques - push and pull - are possible. Additionally, mDNS-SD is already in widespread use in a wide variety of devices and programs.

The service protocols of IP networks are not usable in WSNs, because of their different properties. The radio link in WSNs is usually of low bandwidth, the transmitted packets are limited in size. The typical IP service protocols are not designed with such link layers in mind and usually assume efficient multicasting support. Although they could principally be used on 6LoWPAN IP Wireless Sensor Networks, the radio link would not be used efficiently. Interfacing with these present-day Internet service protocols, however, is needed for the usage of services between WSN and IP networks.

### 2.2.2 Service frameworks in WSN

Some work has previously been done on service discovery in WSNs.

For example in Content-Based Publish Subscribe (CBPS) [Hau05; FHK04] service announcements/requests are attached to the control packets which are used for routing. The service announcements are also cached locally on nodes, along the routing path, to speed up the service for future requests. This approach follows the generic approach of reactive routing protocols with the addition of temporary caching of routes. The Service Request Extension (SRE) and Service Reply Extension (SRpE) are attached in front of the routing control packets. This approach is not possible across network borders and is not based on standardised routing protocols.

A simple services based architecture for a sensor network is proposed in [Gol+03]. A proxy (surrogate host) is used for translating requests from standardised protocols to proprietary protocols and for forwarding them to the sensor nodes. The forwarded request is processed by the nodes in a cooperative manner and the reply is sent back. The actual mechanisms of SD are not detailed in [Gol+03], only an outline for a possible implementation strategy is provided.

In Reflective Middleware for WSNs proposed in [Del+05], a software architecture similar to that in [Gol+03] is used and an implementation based on SOAP and XML is proposed. The usability of these protocols is not realistic on resource limited devices and low-bandwidth radio links.

The main focus of the proposed service-oriented architecture TinySOA in [AG07] is to provide end user application programmers with a simple service-oriented Application Programming Interface (API) to access the WSNs with the programming languages of their choice. Services are registered with the specialised node, i.e. the gateway. The architecture proposed is not decentralised and thus not usable across network borders, so that the application domains are limited.

The authors of [RE07] propose a Service-Oriented Sensor-Actuator Network (SOSANET) architecture with an implementation named TinySOA for TinyOS. This TinySOA is different from the one mentioned in [AG07] as this is a distributed architecture in terms of discovering services from end user applications. The service framework is tightly bound to its routing layer, which makes it hard to use it with other routing layers such as RPL.

Table 2.3 summarises the applicability of existing WSN service discovery protocols to the logistics application scenarios in terms of their requirements. From the summary it can be seen that SOSANET can meet most of the requirements but its essential discovery mechanism is based on its routing protocol (Service-Driven Routing Protocol (SDRP)) which is specifically developed for the purpose of discovery and serving requests. This tight integration makes it difficult to implement the service discovery mechanism presented in SOSANET on sensor networks which depend on routing mechanisms provided by the specific underlying operating system. The dependency on specific routing protocols hinders the discovery of services across network borders.

## 2.2 Service Discovery

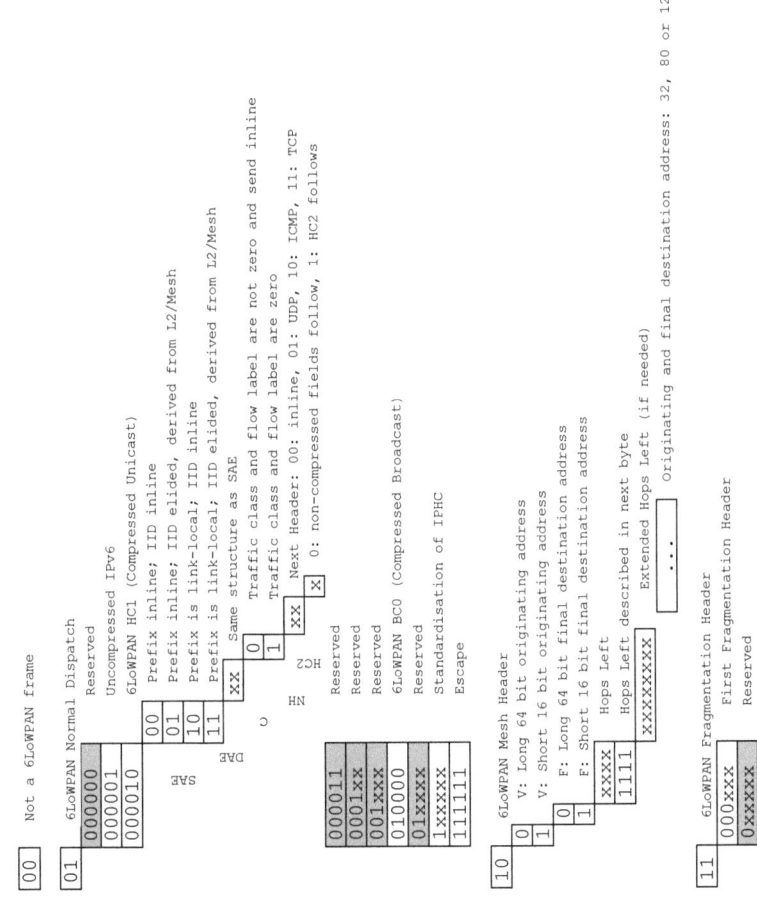

**Figure 2.6: Four 6LoWPAN Dispatch Headers**

Table 2.1: WSN Hardware Platforms

| Company | Memsic | Virtenio | Meshnetics/Atmel |
|---|---|---|---|
| Platform name | TelosB [MEM13] | Preon32 [Vir13] | ZigBit Amp [Atm13] |
| Frequencyband | 2,4 GHz | 2,4 GHz | 2,4 GHz |
| Output power | 0 dBm | 3 dBm | 20 dBm |
| Radio chip | CC2420 | RF231 | RF230 |
| Mikrocontroller | MSP430F1611 | ARM Cortex-3M | Atmega1281 |
| RAM | 10 kB | 64 kB | 8 kB |
| ROM | 48 kB | 256 kB | 128 kB |
| Interface | USB | Micro-USB, JTAG | Mini-USB |

| Company | Zolertia | Intel | Ambient Systems |
|---|---|---|---|
| Platform name | Z1 [Zol13] | iMote2 [Int13] | SmartPoints 3000 [Amb13] |
| Frequencyband | 2,4 GHz | 2,4 GHz | 2,4 GHz |
| Output power | 0dBm | 0dBm | 0dBm |
| Radio chip | CC2420 | CC2420 | CC2430 |
| Mikrocontroller | MSP430F2xxx | PXA271 Xscale | unknown |
| RAM | 8 kB | 256 kB | 8 kB |
| ROM | 92 kB | 32 MB | unknown |
| Interface | Micro-USB | Mini-USB | Pinheader |

## 2.2 Service Discovery

|  | SLP | Jini | UPnP | Bonjour/ mDNS-SD |
|---|---|---|---|---|
| Developed by | IETF | Sun Microsystems | Microsoft | Apple |
| Architecture | Centralised or Distributed | Centralised | Distributed | Distributed |
| Service Registration | Unicast to Directory Agent (DA) or multicast advertisements | Contact lookup service | Multicast advertisements | Multicast advertisements |
| Service Discovery | Unicast to DA multicast to Service Agent (SA) | Query to lookup service | Multicast query | Multicast query |
| Service Technique | Pull only | Pull or Push | Pull or Push | Pull or Push |

**Table 2.2: Service Discovery Protocol Comparison (Extended from [Öst+06])**

| Requirements | CBPS | Reflective Middleware | TinySOA | SOSANET |
|---|---|---|---|---|
| Decentralised | No | No | Yes | Yes |
| Scalable | No | No | Yes | Routing Protoocl Dependent |
| Energy Efficient | Yes | Yes | No | Yes |
| Ease of Service Activation | No | Yes | Yes | Yes |
| Modular Deployment | Yes | Yes | Yes | Yes |
| Small Code Size | Yes | Yes | No | Yes |
| Optimal (low) message overhead | Yes | No | Yes | Routing Protocol Dependent |

**Table 2.3: Service Frameworks for WSNs**

## 2.3 Trickle Algorithm

The Trickle algorithm has initially been introduced to distribute firmware to all nodes in a network efficiently in [Lev+04]. It tries to create consistency of information in a distributed network. The definition of consistency is left to the user of the algorithm. It rapidly propagates information into the network, when new information is available, while limiting the overhead, when no new information needs to be sent. The behaviour of Trickle is to broadcast information at randomly chosen times when few other nodes in radio range have sent the same information in a past time interval. The random time chosen is from an interval that increases, when no new information is detected. When new information is detected, the interval will be reset to a low value. The algorithm scales for a wide variety in node densities. It sends only few packets per hour (when no change is detected) and propagates quickly (when a change is detected). These features make the algorithm a good candidate for the distribution of service messages in networks.

Currently, the algorithm is employed in the routing protocol RPL [Win+12] for the distribution of DIOs. Because of the general applicability of the algorithm, the definition of Trickle has been extracted from the RPL specification into its own IETF RFC 6202 [Lor+11] so that future protocols can integrate it and reference it.

### 2.3.1 Trickle Variables

The variables used by the Trickle algorithm are:

$\tau$ communication interval length (in seconds); in RFC 6206: $I$

$T$ timer value in range $[\tau/2, \tau]$ (in seconds)

$C$ communication counter (without unit)

$\tau$ is an exponentially increased interval length, if the received information from the node's neighbouring nodes is consistent, otherwise it is reset to a low value. $T$ is chosen in the second half of $\tau$, to allow for non-synchronised operation of Trickle. If $T$ would be chosen from the full length, instabilities might appear in a non-synchronised network, cmp. [Lev+04]. The counter $C$ counts how often a node has received consistent data from its neighbouring nodes.

A single Trickle communication interval or Trickle period is shown in figure 2.7. The Trickle period consists of the RX half period (shown as the shaded reactangle) and the TX half period. One specific timer value is shown by the vertical bar.

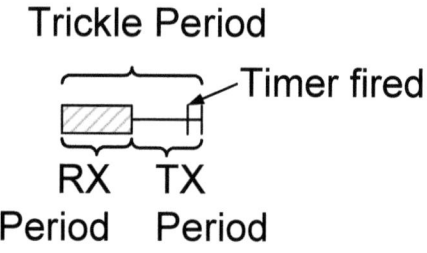

**Figure 2.7: Trickle Period**

### 2.3.2 Trickle Constants

The constants that configure Trickle's behaviour are:

$K$  redundancy constant (typically set to a low integer value, without unit)

$\tau_L$  lowest $\tau$ (in seconds); in RFC 6206: $Imin$

$\tau_H$  highest $\tau$ (in seconds); in RFC 6206: $Imax$

The constant $K$ configures how often consistent information needs to be received by a node from neighbour nodes, so that that node does not transmit. $\tau_L$ denotes the lowest $\tau$, which is in effect, when inconsistencies are detected. It thus governs how much traffic is generated to disseminate new information and to eliminate inconsistencies. $\tau_H$ gives the longest possible interval, which is used in the steady state when all consistencies have been eliminated. Thus, $\tau_H$ governs how much traffic is generated in the consistent case, so that newly arriving nodes can still be updated.

The exponential increase of $\tau$ from $\tau_L = 2s$ to $\tau_H = 32s$ is depicted in figure 2.8. The specific values of $2s$ and $32s$ have been chosen as to meet typical application demands in logistics. The results shown later are based on these values as well, but can be scaled to different $\tau_L$ and $\tau_H$ easily as described in section 9.1.

### 2.3.3 Trickle Rules

The behaviour of Trickle is governed by the following four basic rules:

$\tau$ **expires**  Reset $C$ to 0, double $\tau$, up to $\tau_H$, pick a new $T$ from range $[\tau/2, \tau]$

$T$ **expires**  If $C < K$, transmit

## 2.3 Trickle Algorithm

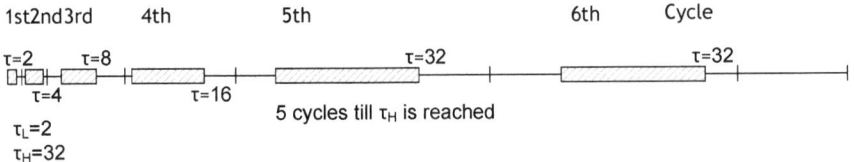

**Figure 2.8: Several Trickle Periods**

**Received consistent data** Increment $C$

**Received inconsistent data** Set $\tau$ to $\tau_L$. Reset $C$ to 0, pick a new $T$ from $[\tau/2, \tau)$

### 2.3.4 Trickle Applicability

The distribution of services in a WSN can also be seen as a consistency problem, where the Trickle algorithm can be applied. The Trickle algorithm does not support the detection of vanished services and the purging of those service information objects from the network though. The integration of this functionality is shown in section 3.6.

In the process of the creation of this thesis a simulation environment for the Trickle algorithm has been created and set up. This environment and simulations of the Trickle algorithm will be presented in chapter 6. Furthermore, analytical models of the Trickle algorithm for its delay and sending behaviour have been derived, which have not been created and published before by other researchers. The models are detailed in chapter 7.

# 3 Service Distribution Algorithms

## 3.1 Requirements of a Service Distribution Algorithm

In order to devise a service distribution algorithm (which also detects vanished services and purges them) that is appropriate for the application domain, requirements for such an algorithm are collected and listed. The algorithm shall meet the following requirements.

Requirements for Distribution:

**R.1** Distribute service to all nodes reliably

**R.2** Distribute service fast

**R.3** Distribute service efficiently

Requirements for Vanish Detection and Purging:

**R.4** Purge vanished service from all nodes reliably

**R.5** Purge vanished service fast

**R.6** Keep silent after purging (efficiency)

**R.6** No false purging service of non-vanished services

**R.7** Limit re-sending of purge by other nodes

General Requirement:

**R.8** Frequency of sending should be uniform across all nodes

## 3.2 Flooding

Flooding is an algorithm which triggers each node to send at least one message on reception of a message [LK01]. In a network of $n$ nodes, there would be thus at least $n$ messages. Flooding is one algorithm that can be used to regularly push services.

The Flooding algorithm has deterministic algorithmic behaviour and provides for a fast dissemination of the service. The algorithm is bound to have collisions, because multiple nodes will retransmit the received message on reception, if there are no probabilistic measures taken against it.

The algorithm is not adaptive with regard to the density of the network. Each node will retransmit a received message although multiple nodes in radio range might already have forwarded the message. In highly inter-connected networks many redundant transmissions are performed.

At deployment time a tradeoff between the frequency of initiating a transmission and the delay of discovering a service needs to be found. There is no adaptivity to the churn[1] of services in the network, so that if a lot of churn is present in the network, the frequency is probably not high enough. If on the contrary the network and its services are stable, the frequency of sending is higher than necessary.

## 3.3 Fixed Interval Pushing

With the algorithm Fixed Interval Pushing each node sends its service table at regular time intervals (denoted by $t_{push}$). On reception of a service packet, each node updates its service table with the received information. With a low setting for the interval services are distributed quickly with a high amount of control packets, while with a high setting for the interval a lower amount of control packets is sent at the cost of a larger delay in service distribution. It is to be noted that the algorithm is not density aware and is thus prone to link-layer collisions in high density networks.

## 3.4 Fixed Inteval Pushing with Vanish Support

The Fixed Interval Pushing algorithm per se does not support the detection of vanishing services. it can be easily extended to detect the vanishing by monitoring, when a service has been last received. With a parameter Time to Live (TTL) it can be configured to how much time has to have passed until a service is assumed to be unavailable. The parameter TTL should be a multitude (e.g. 2−5 times) of the push interval $t_{push}$, so as not to purge services that just have not been received in the last interval.

---

[1] In the Wireless Sensor Networks research area the term churn usually denotes the arrivals and departures of nodes. The churn rate is the number of arrivals and departures in a certain duration. In the thesis a service churn rate is evaluated.

## 3.5 Trickle

The algorithm Trickle has been described in the State of the Art in section 2.3. The algorithm probabilistically distributes information into a network in a density and churn adaptive manner.

The algorithm does not include a mechanism to remove service information from a network, if the provider is removed from the network. The algorithm would indefinitely continue to distribute the information. In the following section solutions to the vanishing of services are discussed.

The vanishing of a service could also be trickled into the network, when detected. However, in the long-term packets are sent for services which are not existing anymore (R.6). If the Trickle algorithm would then be just stopped after some time to stop further sending, no guarantee could be given that the service is purged on all nodes (R.4).

## 3.6 Trickle with Vanish Support

An extension to the original Trickle algorithm to support the vanishing of a service without a need for dedicated deregistration os the service at other nodes is proposed.

The Trickle variables are extended with the following variables:

$V$     Version number for vanish detection

$C_V$     Version counter

The Trickle constants are amended with:

$V_{missed}$ Maximum missed version numbers

The behaviour of Trickle is now additionally governed by the following rules:

- The originating sender increases the version number $V$ every time it sends and includes it in the transmitted packet.
- Non-originating nodes are
    - incrementing $C_V$, when the received version number is the same as the stored version number $V_{RX} = V$
    - setting $C_V = 0$ and $V = V_{RX}$, when the received version number is bigger than the stored version number $V_{RX} > V$

- ignoring the packet, when the received version number is smaller than the stored version number $V_{RX} < V$

- The new $T$ is picked from the modified interval $[1 - \frac{1}{2^{C_V+1}}\tau, \tau)$. The modified interval ensures that nodes with lower $V$ values, which also denotes more recent information, tend to send more often compared to nodes which have higher $V$ values.

- When $C_V > V_{missed}$ the service is marked as vanished and not distributed anymore.

## 3.7 Algorithm Comparison

Table 3.1 summarises the properties of the algorithms described in this chapter.

Flooding has not be studied further in the simulations, analytical models and measurements of this thesis because of its known definciency of broadcast storms. The Pushing and Trickle algorithms have been studied with their extensions to support the detection of vanished services in order to evaluate the fulfillment of the requirements laid out in this chapter.

While the original Trickle algorithm requires 3 variables per service, the modified Trickle algorithm requires 5 variables per service. The Pushing algorithm only needs one variable for a timer per service.

## 3.7 Algorithm Comparison

|   | Flooding | Pushing | Trickle |
|---|---|---|---|
| + | fast dissemination<br>support for vanish | ease of implementation<br>bounded delay | density adaptive<br>churn adaptive<br>no tradeoff |
| − | collisions | not density adaptive<br>tradeoff frequency ↔ delay<br>no support for vanish per se | more complex algorithm |

**Table 3.1: Comparison of Algorithm Properties**

# 4 Wireless Sensor Services Network Framework

Services in information technology are a well-known concept, for example the services that lower-layers are providing to the higher layers in the International Organization for Standardization (ISO)/Open Systems Interconnection (OSI) reference model or web services or application layer service protocols.

Services in IP based networks have been a research focus for quite a while and several standards for the announcement and discovery of services have been published and developed as shown in section 2.2.2. Services are appealing as well as a way to put Autonomic Communication and Self-* algorithms to reality, because the distributed offering of services allows for autonomous and robust service composition.

In figure 4.1 a *data storage sink service*, which allows other nodes to store their generated data, is offered in an IP network, which is connected to a WSN by an IP host with a sensor node attached. If this host would be able to forward services across the boundary, the WSN node, which needs a data storage service, could discover the service from the IP network.

Another application is depicted in figure 4.2, where a node on the IP network is looking for readings from a specific sensor. It could take advantage of the service provided by a node offering this service in the WSN, if the gateway node/host combination would be able to forward service announcements across the network boundary.

When devising a service solution for WSNs, the intrinsically decentralised nature of WSNs has to be taken into account. The service protocol should also be decentralised. Nodes are failing spontaneously, which would mean that, if a centralised service directory is used and fails, the whole service network is failing. A decentralised service protocol would survive the failure of individual nodes.

In order to create an implementation of a service framework, open source implementations of widely used protocols are preferable as those allow for introspection of the code and ease the development of the service framework.

Having evaluated the available IP network service solutions presented in the state of the art in section 2.2, the decision was taken to base the WSN service solution presented here on the mDNS-SD Bonjour protocol. It supports decentralised operation and there is a possibility to have several gateways between the WSN and IP networks. The content of the service messages is encoded in a simple format

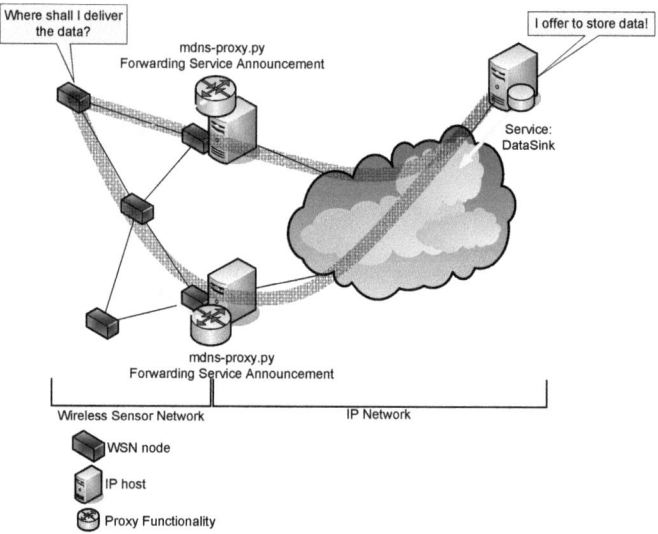

**Figure 4.1: Discovering Internet Services in the WSN**

and can be deciphered even by sensor nodes. Additionally, mDNS-SD is already in widespread use in a wide variety of devices and programs. The implementation of mDNS-SD, used on the IP network side, is Avahi [Poe06]; it offers bindings to many programming languages for interfacing with the service application.

The objective of the service framework is to enable service discovery in WSNs and across network boundaries between Wireless Sensor Network (WSN) and Internet Protocol (IP) based networks. To fulfill this objective the components with the following functionalities had to be implemented or extended in TinyOS 2:

1. Service functionality on the WSN node: announcing or requesting services, caching local and remote services, purging of expired services based on the TinyOS application UDPEcho

2. Service PppRouter functionality on the WSN/Internet Gateway node: Forwarding Service Announcements over the USB interface between the node and the host computer

3. Service Proxy on the WSN/Internet Gateway host: Forwarding Service Announcement between the host computer and the WSN node over USB

4. Zero Configuration enabled applications on Internet hosts implementing

## 4.1 Service Layer Software Components

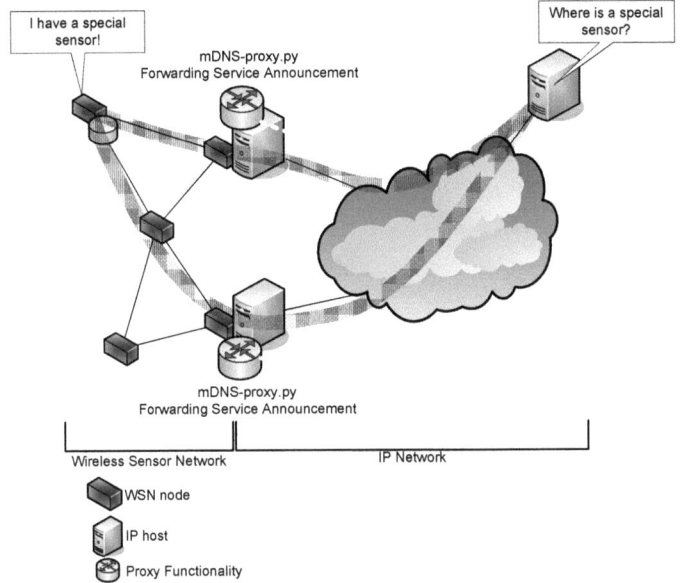

**Figure 4.2: Discovering WSN Services in the Internet**

the functionality 'ZeroConf Service Announcement' and 'ZeroConf Service Discovery'

The protocol stack of the solution is shown in figure 4.3. The lower layers are provided by the TinyOS 6LoWPAN implementation blip-2.0 and the RPL implementation TinyRPL. The Service Proxy on the Border Router node translates between multicast and unicast addresses as detailed in [Ded12].

The interfaces offered and used by the Service layer are shown in figure 4.4. The Application can use the interface ServiceInterface (detailed in algorithm 4.1) to add and remove local services to and from the service layer and be informed about remote services. The Service layer uses the UDP interface, which is offered by blip.

## 4.1 Service Layer Software Components

A TinyOS component diagram of a component using the service layer is shown in figure 4.5. The component includes the MainC component (for booting), the

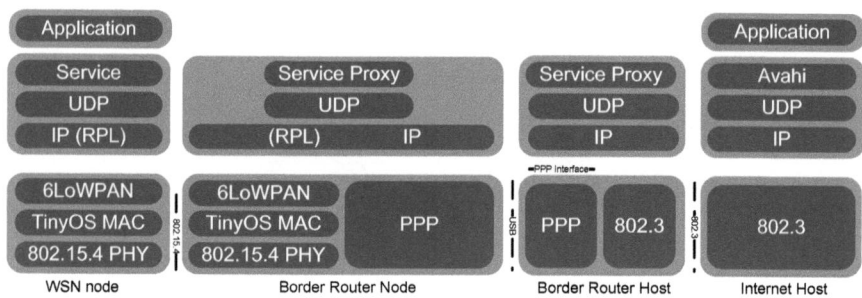

**Figure 4.3: Services Framework Protocol Stack**

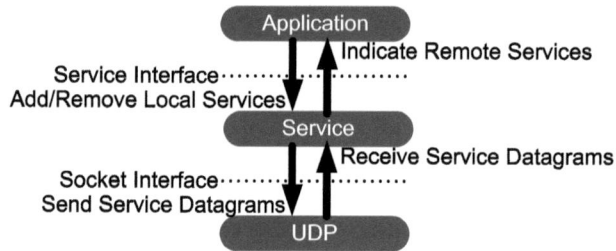

**Figure 4.4: Services Framework Interfaces**

IPStackC (for initialising the IP stack), and the service layer component ServicemdnsC. The Timer component is a generic component (marked with at dashed box), which means that it needs to be instantiated with a parameter (here the resolution of the timer).

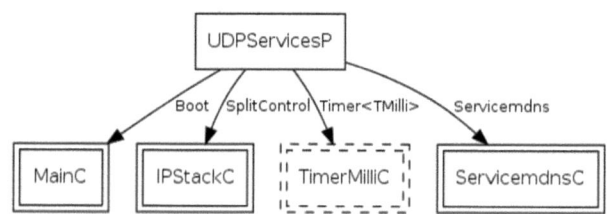

**Figure 4.5: Software Component Diagram: Service User**

A TinyOS component diagram of the component offering the service layer is shown in figure 4.6. The component includes the LocalTimeMilliC component

## 4.2 Service Frame Format

```
interface ServiceInterface {
  command error_t init();

  command error_t add_service(char * service_name,
                              uint8_t service_len);
  command error_t remove_service(char * service_name,
                                 uint8_t service_len);
  command error_t find_service(char * service_name,
                               uint8_t service_len,
                               struct sockaddr_in6 *
                                   measurement_dest);

  command error_t send_service(uint8_t id);

  event void service_changed(uint8_t id);
  event void service_vanished(uint8_t id);
}
```

**Algorithm 4.1: Service Interface**


(for various timers needed for local and remote services) and the UDPSocketC (for UDP based distribution of the services).

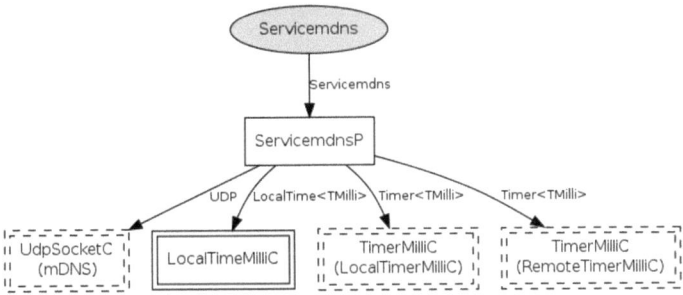

**Figure 4.6: Software Component Diagram: Service Layer**

## 4.2 Service Frame Format

The frame formats for the service announcements are shown in figure 4.7 and 4.8. The VERSION field contains the version of the protocol and is set to 0. The short frame format is used, when the service is located in the WSN as then only the 16 bit node ID needs to be transmitted. For services in the IP network, the full IPv6 address needs to be included in the frame as SENDER. Which of the messages is in use, is denoted in the FLAGS field. DEFAULT_PORT is the UDP port on which the service is available, when no port is given in the TXT description of the service. The TRICKLE_VERSION is used to support the detection of the vanish of a service. Only the injecting node is allowed to change the version number. Non-injecting nodes can monitor the update of the version number and purge the service from the local service cache if it hasn't changed for some time. The HOPS field is set to the minimum received HOPS field for that service plus one when distributing the service. It is set to zero by the injecting node. The TXT field is used for the actual service description as specified in [CK13a]. The TXT_LEN field gives the length of the TXT field.

```
 0                   1                   2                   3
 0 1 2 3 4 5 6 7 8 9 0 1 2 3 4 5 6 7 8 9 0 1 2 3 4 5 6 7 8 9 0 1
+-+-+-+-+-+-+-+-+-+-+-+-+-+-+-+-+-+-+-+-+-+-+-+-+-+-+-+-+-+-+-+-+
|     VERSION     |     FLAGS       |            SENDER         |
+-+-+-+-+-+-+-+-+-+-+-+-+-+-+-+-+-+-+-+-+-+-+-+-+-+-+-+-+-+-+-+-+
|         DEFAULT_PORT              |            TTL            |
+-+-+-+-+-+-+-+-+-+-+-+-+-+-+-+-+-+-+-+-+-+-+-+-+-+-+-+-+-+-+-+-+
|        TRICKLE_VERSION            |   HOPS    |    TXT_LEN    |
+-+-+-+-+-+-+-+-+-+-+-+-+-+-+-+-+-+-+-+-+-+-+-+-+-+-+-+-+-+-+-+-+
|   TXT ...
+-+-+-+-+-+-+-+-+-+-+-+-+-+-+-+-+-+-+-+-+-+-+-+-+-+-+-+-+-+-+-+-+
```

**Figure 4.7: Frame format (Short)**

```
 0                   1                   2                   3
 0 1 2 3 4 5 6 7 8 9 0 1 2 3 4 5 6 7 8 9 0 1 2 3 4 5 6 7 8 9 0 1
+-+-+-+-+-+-+-+-+-+-+-+-+-+-+-+-+-+-+-+-+-+-+-+-+-+-+-+-+-+-+-+-+
|     VERSION     |     FLAGS       |            SENDER         |
+-+-+-+-+-+-+-+-+-+-+-+-+-+-+-+-+-+-+-+-+-+-+-+-+-+-+-+-+-+-+-+-+
|                              SENDER                           |
+-+-+-+-+-+-+-+-+-+-+-+-+-+-+-+-+-+-+-+-+-+-+-+-+-+-+-+-+-+-+-+-+
|                              SENDER                           |
+-+-+-+-+-+-+-+-+-+-+-+-+-+-+-+-+-+-+-+-+-+-+-+-+-+-+-+-+-+-+-+-+
|                              SENDER                           |
+-+-+-+-+-+-+-+-+-+-+-+-+-+-+-+-+-+-+-+-+-+-+-+-+-+-+-+-+-+-+-+-+
|            SENDER             |          DEFAULT_PORT         |
+-+-+-+-+-+-+-+-+-+-+-+-+-+-+-+-+-+-+-+-+-+-+-+-+-+-+-+-+-+-+-+-+
|               TTL             |         TRICKLE_VERSION       |
+-+-+-+-+-+-+-+-+-+-+-+-+-+-+-+-+-+-+-+-+-+-+-+-+-+-+-+-+-+-+-+-+
|     HOPS      |    TXT_LEN    |     TXT ...
+-+-+-+-+-+-+-+-+-+-+-+-+-+-+-+-+-+-+-+-+-+-+-+-+-+-+-+-+-+-+-+-+
```

**Figure 4.8: Frame format (Long)**

## 4.3 Service Forwarding in the Wireless Sensor Network

When trying to distribute information in an 6LoWPAN to a multitude of nodes an IPv6 address is needed. The possible multicast network prefixes for the first 64 bit of the IPv6 address are shown in table 4.1. In the 6LoWPAN implementation blip-2.0 the ff02:: prefix is used for radio-range, while the ff05:: prefix is routed to the complete 6LoWPAN.

| Prefix | Description |
|---|---|
| ff01:: | Interface-local, same node. Similar to loopback addresses |
| ff02:: | Link-local, same link (radio range in blip-2.0) |
| ff03:: | IPv4 local scope |
| ff04:: | Admin-local |
| ff05:: | Site-local, same site (routed in blip-2.0) |
| ff0e:: | Global scope, eligible to route via the Internet |

**Table 4.1: Multicast Network Prefixes [Source: [Ded12]]**

The second half of the IPv6 address is the interface identifier. Possible values are listed in table 4.2. For Multicast Domain Name System for IPv6 (mDNSv6) the interface identifier ::fb is reserved by [CK13b].

Since Trickle is used for the distribution and not the routing protocol flooding, the IPv6 address that has been used is ff02::fb.

| Interface Identifier | Description |
|---|---|
| ::01 | All nodes |
| ::02 | All routers |
| ::fb | mDNSv6 |
| ::1:2 | All Dynamic Host Configuration Protocol (DHCP) servers and agents |
| ::101 | Network Time Protocol |
| ::108 | Network Information Service |

**Table 4.2: Multicast Interface Identifier [Source: [Ded12]]**

## 4.4 Internet Host Service Application

Figure 4.9 shows the graphical user interface of 'avahi-discover', which lists the services of type '_wsn._udp': Temperature, Light, Humidity. The services are provided by the TelosB motes in the network. The service type '_wsn._udp' stands for WSN and UDP. Additionally a database service ('WSN DB') from the IP network is shown, which can be used by the nodes to permanently store measured data. This implementation has been used in multiple technical demonstrations.

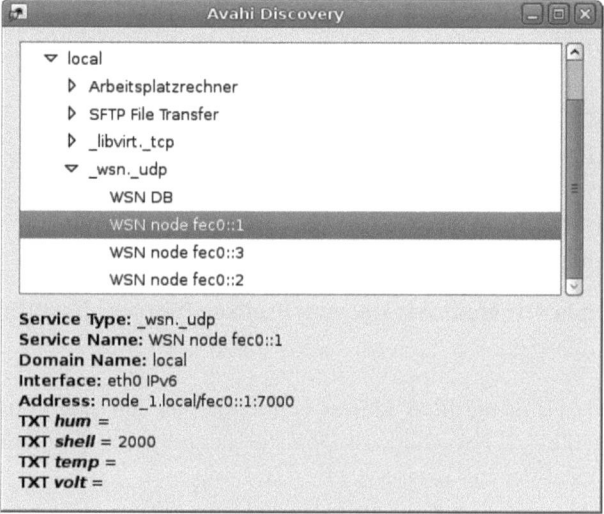

**Figure 4.9: Avahi Integration**

# 5 Evaluation Metrics and Scenarios

The algorithms of the service framework have to be evaluated. Therefore a number of evaluation metrics have been chosen (section 5.1). Additionally, various scenarios have been set up to cover the most common network topologies (section 5.2).

## 5.1 Evaluation Metrics

The next sections are giving an overview of the evaluation metrics *Number of Packets Sent*, *Energy Spent*, *Time to Consistency*, and *Scalability*. These metrics have been selected because they determine whether the algorithms can be used effectively, efficiently and reliably (cf. the requirements in section 3.1). The metrics will be used for all three evaluation methods (simulation, analytical modelling and measurements).

### 5.1.1 Number of Packets Sent

The main metric to evaluate is the number of packets that are sent by the algorithm within a certain period. Since the Trickle algorithm adapts its rate over time, the transmission rates are not static, but depend on the observation duration. Therefore a specific duration is chosen and the number of packets is counted and compared between the algorithms. The duration should be a multitude of $\tau_H$, so that the Trickle algorithm can adapt its rate to the lowest rate within the observation duration. For most results in this thesis $\tau_H$ has been set to $32s$ and an observation time of $400s$ was allowing for the Trickle algorithm to adapt to its lowest rate and distribute to most nodes in all scenarios.

The number of packets can be evaluated on a per node base as well as for the complete network. The number of packets sent for the complete network is denoted by $M$ (the mean number of packets for the complete network from several Monte-Carlo runs is denoted by $\overline{M}$), while the number of packets of node $i$ is denoted as $M_i^*$. The mean number of packets per node is denoted by $\overline{M^*}$. A minimisation of the complete packet number is desirable, under the constraint of an almost uniform distribution of the messaging among the nodes. It might however also be desirable that nodes closer to the source of the service send more packets than nodes further away, in order to allow for a spatial restriction of a service. Furthermore, the metric

*Number of Packets* has to be traded-off against the metric *Time to Consistency* (section 5.1.3). A low number of packets comes with a higher time to consistency.

### 5.1.2 Energy Spent

The energy spent by the algorithm is another important metric to evaluate. Since the energy spent on computation is neglegible compared to the energy used for communication by transmitting radio packets [Beh09], this value can be derived mostly from the number of packets sent. Again, this can be evaluated on a per node (denoted by $E_i^*$) or in total (denoted by $E$). The mean energy spent per node is denoted by $\overline{E^*}$. Since the energy spent can be derived from the number of sent packets (e.g. using the energy information from [Bou+08]), this metric will not be shown explicitly in this thesis.

### 5.1.3 Time to Consistency

As services should also be effectively distributed, the time until nodes have reached consistency is a major metric which needs to be evaluated. The time to consistency can be described by the Probability Density Function (pdf) or the Cumulative Distribution Function (cdf) for the individual nodes or the complete network. The time is denoted by $t$, the pdf is denoted by $p(t)$, while the cdf is denoted by $P(T \leq t)$. It should be pointed out again that *Time to Consistency* needs to be trade off against the *Number of Packets Sent* to meet the application requirements.

### 5.1.4 Scalability with the Number of Nodes

Since the algorithms have to scale for networks of various sizes, it is important to evaluate them for the behaviour with an increasing number of nodes. Preferably the number of messages $M$ should increase less than linearly with an increasing number of nodes ($< \mathcal{O}(N)$).

### 5.1.5 Scalability with the Number of Services

When the number of services in the network or per node increases, this should lead to a less than linear increase in the number of messages $M$ ($< \mathcal{O}(S)$).

## 5.2 Scenarios

In order to study the behaviour of the service algorithms, several scenarios have been set up to cover the most common network topologies. These scenarios will

## 5.2 Scenarios

be described in the following sections. The parameters, which can be controlled for the scenarios are the number of nodes in the scenario and the distance between the nodes. In the following the scenarios will exemplary be depicted for 9 nodes and, if appropriate, 10 m inter-node distance.

### 5.2.1 Line Scenario

In this type of scenario, the topology consists of all nodes being arranged on only one axis in a line. This type of scenario could e.g. be used to derive results for street lighting applications.

#### 5.2.1.1 Only Direct Neighbours

In this sub-type of the Line scenario, the nodes are only connected to their direct neighbours to the left and right (if available). The pathloss between neighbouring nodes is governed by the standard TinyOS Simulator's propagation model CPM [LCL07], the resulting PRR topology is shown in figure 5.1. This scenario is not realistic, but gives insights for the creation of the analytical model. This scenario will be referenced by the name 'Line-Direct'.

#### 5.2.1.2 Neighbours According to Propagation Model

In this sub-type of the Line scenario, all nodes and not just the neighbouring nodes are connected according to the CPM propagation model, as shown in figure 5.2. It therefore resembles propagation in WSNs more realistically. This scenario will be referenced by the name 'Line-CPM'.

### 5.2.2 Grid Scenario

The nodes are aligned regularly in a grid. Each node has the same distance to its closest neighbours. This scenario will be referenced by the name 'Grid-CPM' and is shown for 9 nodes in figure 5.3. This type of scenario could be used to derive results for wild-life monitoring and other rectangular area spanning applications. Note that at a low distance of 10 m the topologies that result from Line-CPM and Grid-CPM are the same, as all nodes are connected to each other by perfect links.

### 5.2.3 Random Scenario

The nodes are placed randomly. Each node's x and y location is drawn from a uniform pseudo random number generator. This scenario will be referenced by

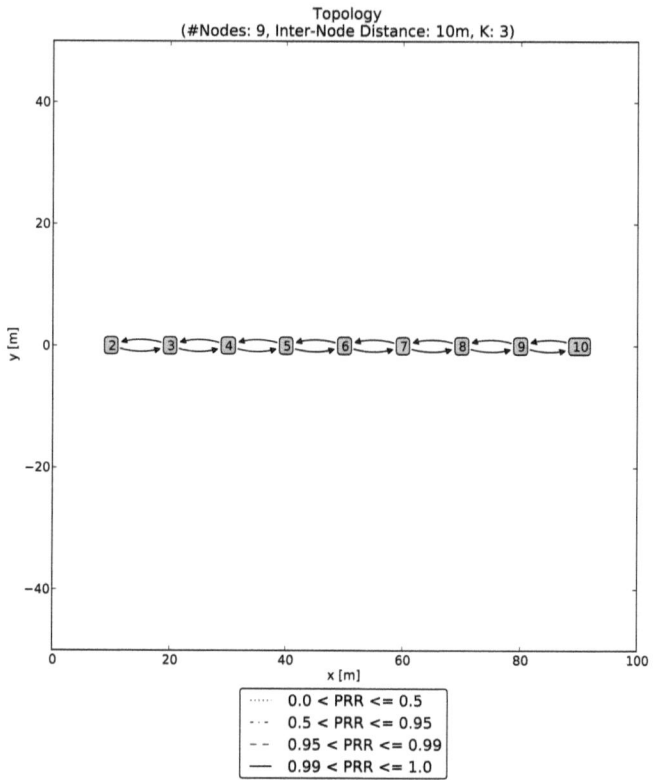

**Figure 5.1: Line-Direct Scenario with 9 Nodes**

the name 'Random-CPM'. One instance of such a scenario is shown in figure 5.4. The dimensions of the scenario are chosen to cover the same area as the Grid-CPM scenario (same 'virtual inter-node distance'). This type of scenario does not exhibit regular structure and can thus minimise the effect of a regular structure on the results. It should be noted that it is possible that individual nodes might be not in coverage of the remaining nodes in individual scenario instances and thus might not necessarily be informed about services.

### 5.2.4 Container Scenario

17 nodes are located at positions that were used for measurements in a cargo container. The scenario is shown in figure 5.5. The propagation conditions on the links

## 5.2 Scenarios

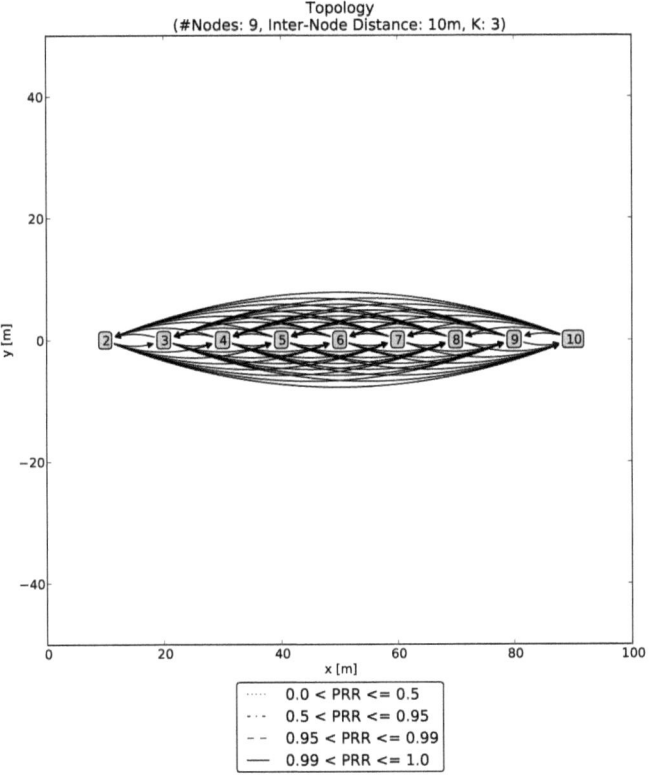

Figure 5.2: Line Scenario with 9 Nodes

between the nodes have been adapted to meet the propagation conditions that were experienced in measurements in a cargo container.

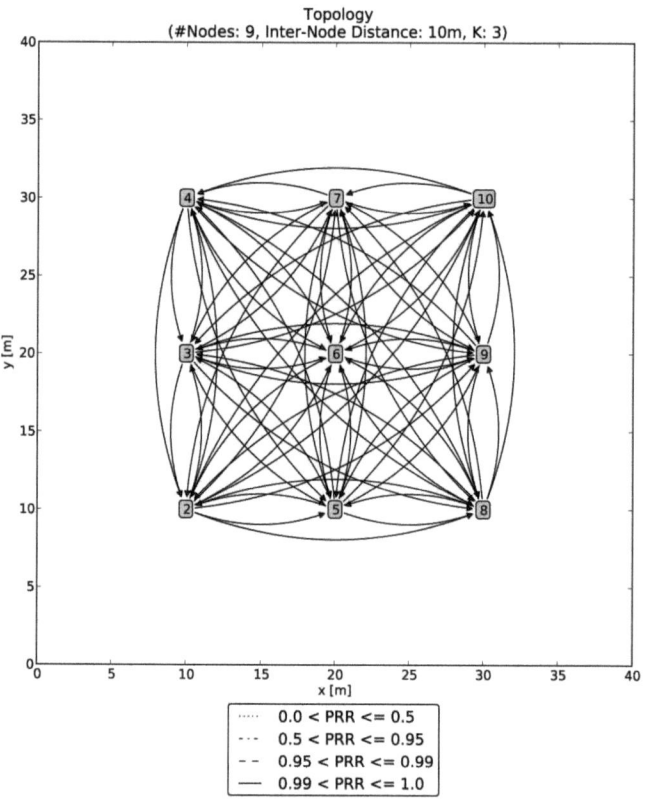

**Figure 5.3: Grid Scenario with 9 Nodes**

5.2 Scenarios

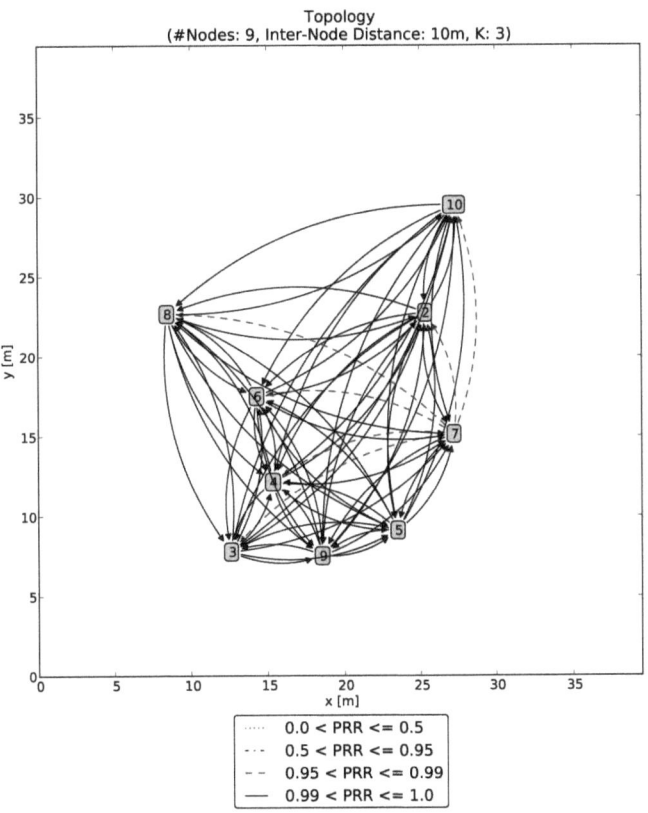

**Figure 5.4: Exemplary Random Scenario with 9 Nodes**

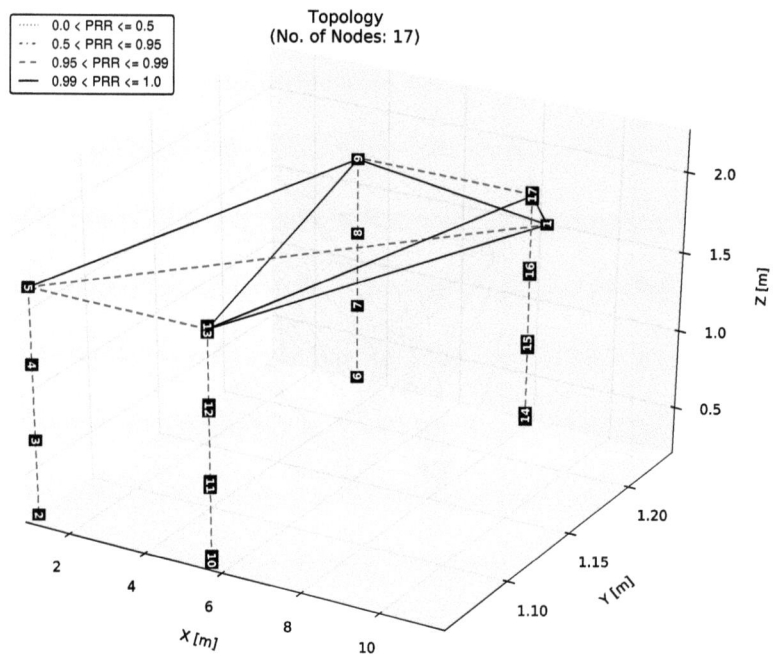

**Figure 5.5: Container Scenario**

# 6 Simulation of Service Discovery

The performance of the Service Discovery algorithms is evaluated by means of simulations, analytical modelling, and measurements. The following sections describe the simulation environment, that has been set up, and shows results which were obtained from the simulations.

## 6.1 Simulation Environment

In order to validate the analytical model, a simulation environment has been setup. The TinyOS simulation tool TOSSIM [Lev+03] has been used in combination with an implementation of the Trickle algorithm in the application layer. The lower layers conform to the Berkeley Low-Power IP (blip) stack, which has been modified in order to be able to simulate it with TOSSIM. Blip's built-in Trickle timer in its ICMP implementation has not been part of this study, solely the application layer Trickle instance has been evaluated.

The simulations were performed with up to 300 Monte-Carlo repetitions for each scenario instance with varying seeds in the TinyOS executable as well as the Python TOSSIM script. The seed in the TinyOS executable governs the Pseudo-Random Number Generator (PRNG) used by the Trickle algorithm. The seed in the Python TOSSIM script governs the PRNG used to generate the boot time of the simulated nodes as well as the radio propagation model. The seed can be seeded with a true random number by the Linux kernel random number source /dev/urandom. For reproducibility of results the seed is logged and can be re-used in later simulation runs. The simulation tool can be used with the previously mentioned scenarios. A simulation suite run consists for the 'Line', 'Grid' and 'Random' type of scenarios of instances of those scenarios with a varying number of nodes $N$ and inter-node distances $d$. The inter-node distance is the distance between neighboring nodes in Line and Grid scenarios. The Random scenario has a virtual inter-node distance in that it spans the same area as the Grid scenario of the same inter-node distance and number of nodes.

## 6.2 Simulation Evaluation

The evaluation of the simulation results is done by post-processing the logging files which are created by the simulation tool TOSSIM using the Python programming language. The output files are processed by using Regular Expressions (RegExps) to match debugging statements, that have been programmed into the source code of the simulated executable. Several evaluation scripts have been created, that create binary, text or graphical output. The graphical output allows for contour and topology plots, packet scatter plots, pdfs and cdfs. The graphical plots are created using the graphing library `matplotlib`. The evaluation can also be started independently from the simulation execution, so that previous simulation runs can be processed with enhanced postprocessing evaluation scripts without the need to re-run the simulations.

The post-processing can be done on individual runs, on a set of Monte-Carlo runs as well as on a complete suite of simulations (i.e. for different scenarios).

Parts of the results will also be re-used for comparison with results from analytical models and measurements.

## 6.3 Simulation Results

In figure 6.1 the results of a simulation of a Line-CPM scenario with 9 nodes and 50 Monte-Carlo iterations is depicted. A more detailed discussion of the simulation results is done in conjunction with the results of the analytical model in section 9.1.1.

### 6.3.1 Statistical Significance

The simulation results which were obtained in this thesis have been analysed for statistical significance. The 95% confidence intervals were calculated by using the Survival module [The12] of the statistical evaluation tool GNU R [R D12]. This module creates survival curves (ccdf) from a Kaplan-Meier maximum likelihood estimator [KM58]. The variation and the confidence interval is internally estimated by Greenwood's formula, cf. e.g. [KP80] and appendix C.

In figure 6.1 the 95% confidence intervals for the Cumulative Distribution Functions (cdfs) for the various distances are shown as dashed lines giving a corridor around the simulated cdfs. As can be seen from the figure, the corridor is rather narrow, especially for low distances. For distances with high PRR on links between direct neighbours (i.e. 130 m and higher) the corridor is slightly wider ($\approx$ 9%), since fewer samples could be gathered in the depicted model time.

## 6.3 Simulation Results

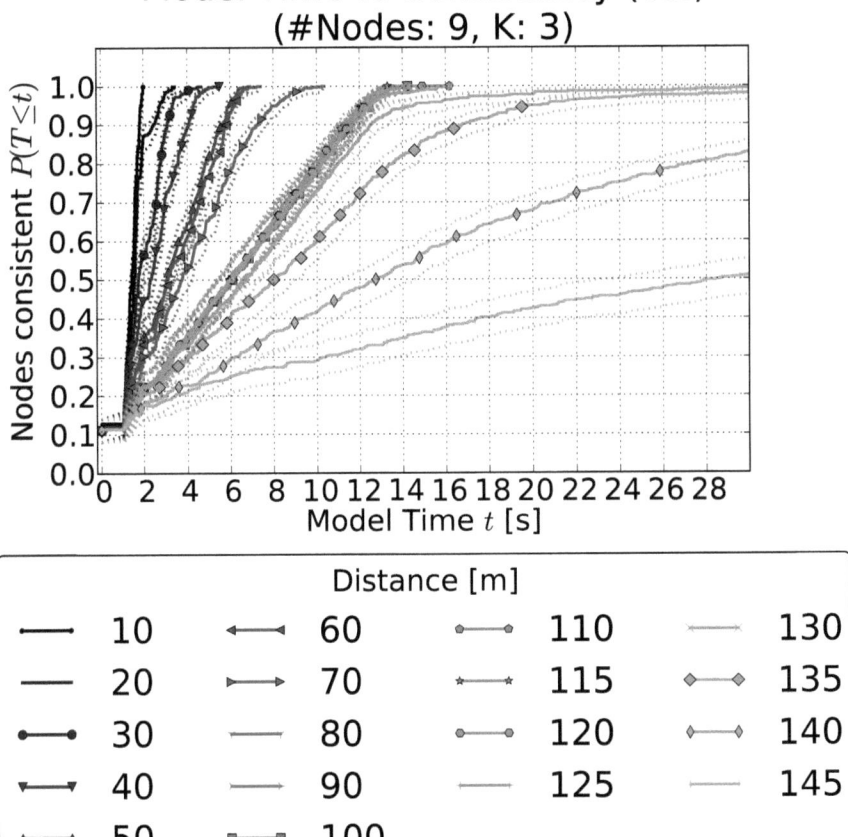

Figure 6.1: Simulation Results for 9 Node Line-CPM Scenario with 95% Confidence Intervals

For scenarios with more nodes (e.g. for a 25 node Line-CPM scenario as shown in figure 6.2) the corridors are even narrower, since more samples can be gathered.

From the width of the 95% confidence interval for the shown scenarios it can be concluded that the execution of 50 Monte-Carlo iterations yields a satisfying statistical confidence in the results.

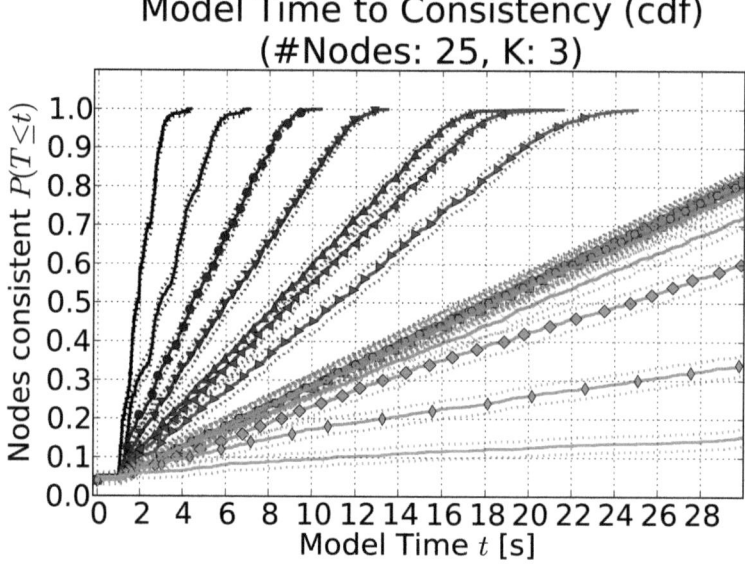

Figure 6.2: Simulation Results for 25 Node Line-CPM Scenario with 95% Confidence Intervals

## 6.3.2 Uniform Spatial Distribution of Sent Packets

The spatial distribution of the number of sent packets is shown in a 100 node Grid-CPM scenario with an inter-node distance of 100 m is shown in figure 6.3. Each grey-scaled rectangle in the graph shows the number of sent packets of the node placed at that location. The number of sent packets varies between 10 and 26 in the observation time of 400 s. The service injection node is on the lower left corner and one service is injected after 10 s. The service gets distributed by the Trickle algorithm and thus is re-sent by nodes. The re-announcing nature of the Trickle algorithm also cares for the distribution of service in dynamic scenarios.

The results in this section are based on the student project thesis of Jens Dede [Ded12] performed at the Communication Networks group of the University of Bremen.

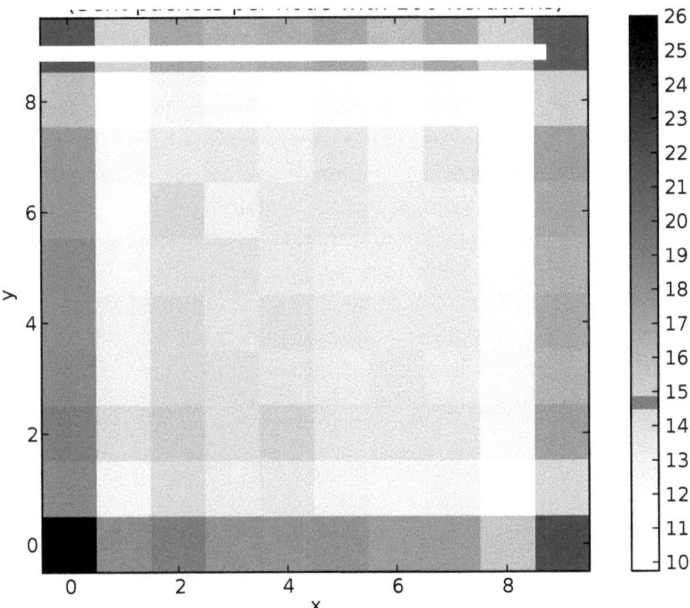

**Figure 6.3: Number of Sent Packets in 100 Node Grid Scenario**
Inter-Node Distance: 100 m, K=3, Observation Duration: 400 s
[Source: [Ded12]]

It can be seen, that the injection node has the highest number of sent packets as it is the one starting the process and thus having the longest active time in the observation period. Additionally, the injection node is one of the corner nodes in the scenario. The corner nodes have only 3/9th of the neighbours compared to the nodes in the center of the scenario. The corner nodes thus show a higher number of sent packets since the Trickle algorithm adapts to the node density. The edge nodes — having 5/9th the number of neighbours of the center nodes — also send more packets compared to the center nodes. Due to the higher number of packets sent by the corner and edge nodes, the nodes in the ring between the center nodes and the outer nodes have lower number of sent packets. The distribution of the number of sent packets in the center of the scenario is almost uniform with a slightly higher number of sent packets when the nodes are closer to the injection node.

### 6.3.3 Results for Grid and Random Scenarios

The difference of the delay cdf between a planned (Grid) and an unplanned random deployment of nodes is shown in figure 6.4 for a 16 node scenario of comparable dimensions. Only a subset of all simulated distances is shown for clarity. From the figure it can be seen that the delay difference between Grid and Random deployment is small for dense deployments. For a virtual inter-node distance (cmp. section 5.2.3) of 80 m the distribution of a service is faster to about 85% of the nodes for the Random topology. For the remaining nodes, which are spaced further apart than in the regular Grid scenario, the distribution is slower with a Random topology. A similar behaviour can be observed for a virtual inter-node-distance of 140 m. Here about 45% of the nodes are informed faster in a Random scenario. However, since the dimensions are large, a big fraction of the nodes is not informed, as the network is separated into clusters, which are not necessarily connected. One random topology of one Monte-Carlo run with one separated node is shown in figure 6.5.

Results comparing Line and Grid scenarios will be shown in section 9.1.1 together with results of the analytical model.

### 6.3.4 Trickle Parameter Analysis

The effect of the Trickle parameters is listed in table 6.1. It can be seen that with reduced $\tau_H$, the steady-state rate is approximately anti-proportionally increased, while no clear effect on the number of packets until steady-state is reached can be seen. An increase of $K$ from 3 to 4 shows up as a slight increase in the number of packets until steady-state is reached as well as in a slightly higher rate in steady-state.

Not shown in the table is that a lower $\tau_L$ leads to a faster distribution and a

## 6.3 Simulation Results

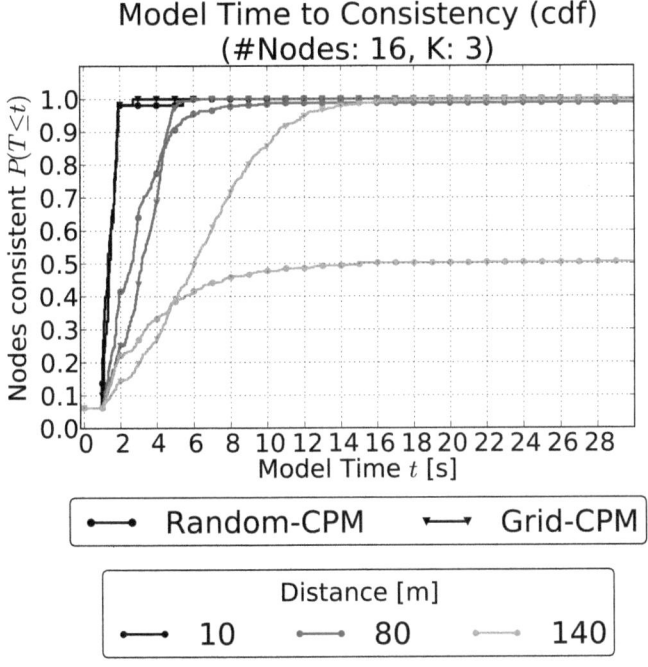

**Figure 6.4:** Network Consistency Delay Distribution (Simulated, 16 nodes)

| $\tau_L$ [s] | $\tau_H$ [s] | $K$ | Packets sent until steady state | Packets sent in steady state [1/s] |
|---|---|---|---|---|
| 2 | 32 | 3 | 124 | 1.8 |
| 2 | 16 | 3 | 123 | 3.6 |
| 2 | 32 | 4 | 141 | 2.1 |
| 2 | 16 | 4 | 140 | 4.2 |
| 2 | 8 | 4 | 140 | 8.6 |
| 2 | 4 | 4 | 170 | 17.7 |

**Table 6.1:** Effect of Trickle Parameters (100 node Grid-CPM, 100 m Inter-Node Distance)
[Source: [Ded12]]

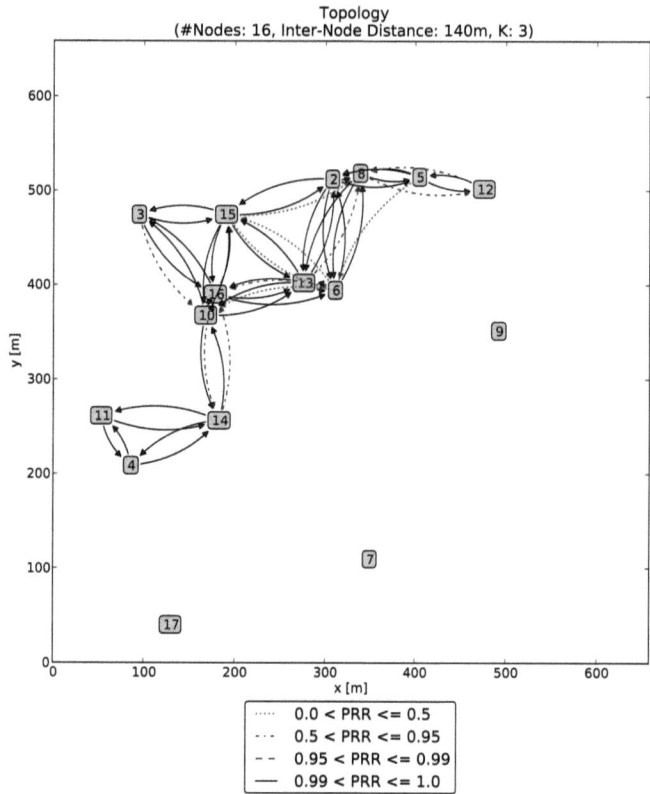

Figure 6.5: 16 Node Random Topology of one Monte Carlo Run

slightly higher sending rate in the beginning and no effect on the steady-state rate. $\tau_L$ is limited on the lower-bound by the link-layer. Especially when the link-layer is duty-cycled, the duty-cycle parameters are limiting the lowest possible value of $\tau_L$.

### 6.3.5 95 Percentiles and Application Requirement Optimisation

The 95 percentiles calculated from the simulation results are tabulated in table 6.2 for the Line scenarios and in table 6.3 for the Grid scenarios for various number of nodes, Trickle K parameters and distances. The 95 percentile gives the time at which 95% of the nodes have been informed about a service ('consistent'). Entries

## 6.3 Simulation Results

marked with '>30' reach the 95th percentile outside the evaluated time, which was set to 30 s in order to have a good resolution of the histogram with reasonable memory requirements.

For the Line as well as the Grid scenarios one can see that the 95 percentile delay increases with the inter-node distance as well as with the number of nodes in the scenario.

The 95th percentile delay of the Line scenario is typically higher than the delay in the Grid scenario as the Grid scenario has more neighbours at each node and a lower scenario diameter. The amount of delays larger than 30 s is thus also larger in the Line scenario.

With the two tables 6.2 and 6.3 application requirements for service discovery (or default route discovery as well) can be brought in accordance with scenario layout (type of scenario, number of nodes in the scenario and inter-node distance). For example it can be derived from the tables, that if a service application requires a service to be found within 2 s with 95% probability, then the scenario should contain 4 nodes separated with 40 m inter-node distance maximum or 9 nodes separated by 10 m only for Line-like scenarios. For Grid-like scenarios, 4 nodes could be spaced by up to 90 m, 9 nodes up to 50 m, 16 nodes up to 30 m, 25 and 36 nodes up to 20 m and at a distance of 10 m a number of nodes up to 100 could be adequately be served.

Similarly, other delay requirements could be deduced from the tables, e.g. for 5, 10 and 30 seconds. Scenarios with a Grid structure can typically support more nodes and higher distances. The results in the tables were obtained for a $\tau_L = 2s$ and are mostly independent of $\tau_H$ when $\tau_H \gg \tau_L$ as Trickle should be parameterised. Estimations for other $\tau_L$ values can be done by linearly scaling the results, e.g. for a $\tau_L$ of 1 s the results in the table should be halved.

Compared to the distance and the number of nodes, the impact of the Trickle parameter $K$ is minor. For low distances and number of nodes, the value of $K$ can be chosen arbitrarily, for higher distances and number of nodes a medium $K$ of 3 seems to be the best option compared to the low and high values. The rather bad delay performance of the high $K$ setting of 9 can be explained that in that case many already informed nodes in the hinterland of the border between consistent and inconsistent nodes are keeping the nodes directly at the border from transmitting.

### 6.3.6 Trickle and Push Comparison

In order to compare the Trickle algorithm with the simple Push algorithm various simulations have been performed with different push intervals. The Trickle parameters used for the simulation results in this section are $K = 3$, $\tau_L = 2$ s, and $\tau_H = 32$ s. The scenario has been a 100 node Grid-CPM scenario with an inter-node distance of 100 m. The results are listed in table 6.4 and shown in a spiderweb chart in figure 6.6.

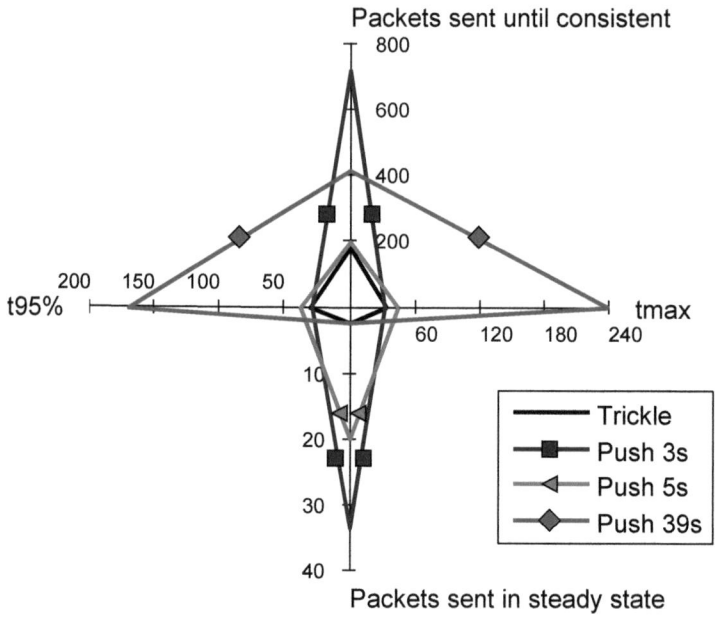

**Figure 6.6: Spiderweb Comparison of the Algorithms [Data Source: [Ded12]]**

It can be seen that for a similar performance to the Trickle algorithm the Push algorithm needs to be configured with a push interval $t_{push} = 3s$. The number of packets sent until the network is consistent by the the push algorithm is 4 times higher compared to the Trickle algorithm. The packet sent rate in steady state of the Push algorithm is more than 10 times as high as Trickle's rate.

With a push interval of $t_{push} = 5s$, the Push algorithm can be configured to have a number of sent packets in the same order of magnitude as the Trickle algorithm, but then the delay performance is worse and the steady-state sending rate is still 8 times higher than the Trickle algorithms' rate.

In order to get a similar steady state rate, the Push algorithm needs to be configured with a push interval of $t_{push} = 39s$. This further degrades the consistency delay metrics $t_{95\%}$ and $t_{max}$. The number of sent packets until consistency is reached within the network increases as well, since parts of the network are sending for a long time, while the remainder of the network is not yet consistent.

The minimum number of packets sent until consistent of the Push algorithm is when it is configured with a medium interval of $t_{push} = 5s$. With a lower setting, packets are sent more frequently, while with a higher setting it takes longer to reach the fully consistent state. The combination of the sending rate and the time until consistency is reached, indicates that there is a minimum number of packets sent at medium push intervals.

### 6.3.7 Service Vanish Comparison

The delay behaviour of the vanish extension to the Trickle algorithm described in section 3.6 is shown in figure 6.7 compared to the behaviour of the Push algorithm. The Push algorithm has been parameterised $t_{push} = 5s$ and a TTL of 20 seconds. The Trickle algorithm has been parameterised with $\tau_H$ (same as $I_{max}$) of 4, 8, and 16 s.

The detection of a new service shows exactly the same behaviour for all three parameterisations of the Trickle algorithm. The Push algorithm is slightly slower to distribute the service.

At $t_{rm} = 122.5s$ the service is removed from the injecting node. The node is still available, but does not actively announce the vanishing of the service. The removal of the service at the injecting node can be seen in the reduced percentage in the figure. Approximately 20 s later the nodes with the Pushing algorithm and a few seconds later the Trickle algorithm with $I_{max} = 4$ start to remove the service from their service caches. The Trickle parameterisations with higher $I_{max}$ show a significant delay in the discovery that the service vanished.

It can be summarised that a high $\tau_H = I_{max}$ which is desired for a low steady-state sending rate, impacts the delay for the discovery of vanishing services. A trade-off between the sending rate and application requirements needs to be performed. It should be noted, that this only effects the incorrect discovery of a vanished services. Nodes actively using a service will discover the absence of the service by non-reactivity of the peer node.

The vanish extension of the Trickle algorithm employs a version number to track the continued presence of services. The progression of the version number can be seen in figure 6.8. The injecting node at the lower left hand corner has the highest version number. In quarter circles around the injecting node the version numbers are showing lower (and thus older) values. When the service disappears, the version number is not increased by the injecting node and the other nodes can detect the static version number and thus the vanishing of the service offering node.

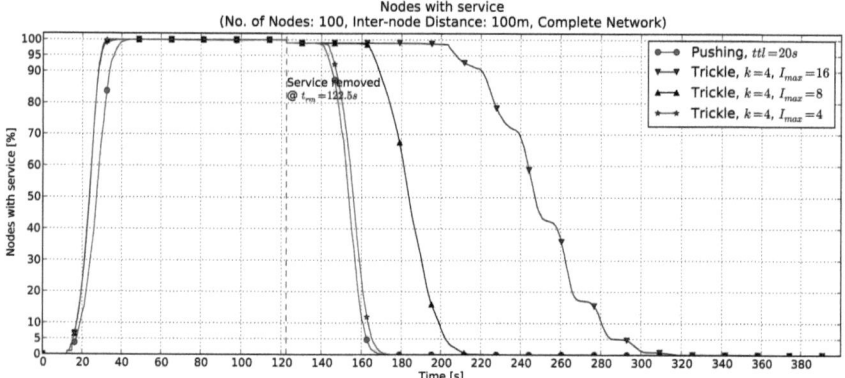

**Figure 6.7:** Comparison of Consistency Delay of Trickle and Push Algorithm [Source: [Ded12]]

### 6.3.8 Routing Protocol Simulation Results

Based on the same simulation tool which is used for the simulation of service discovery, the behaviour of the Trickle algorithm in the routing layer can be evaluated as well. In RPL, DIO messages are used to distribute the information for the creation of default routes (routes to the root node of the network). The transmission of the DIO messages is controlled by the Trickle algorithm. The Trickle algorithm of the routing layer was parameterised as shown in table 6.5. In figure 6.9 the cdf of the time to the discovery of the default route is shown for both standardised RPL objective functions Objective Function 0 (OF0) and Minimum Rank Objective Function with Hysteresis (MRHOF). Both cdfs show a similar behaviour with the linear increase between 128 ms and 256 ms, when compared to the service layer results presented earlier. For more than 256 ms the cdfs follow an s-shaped curve. The results shown here have been obtained using the container scenario.

## 6.3 Simulation Results

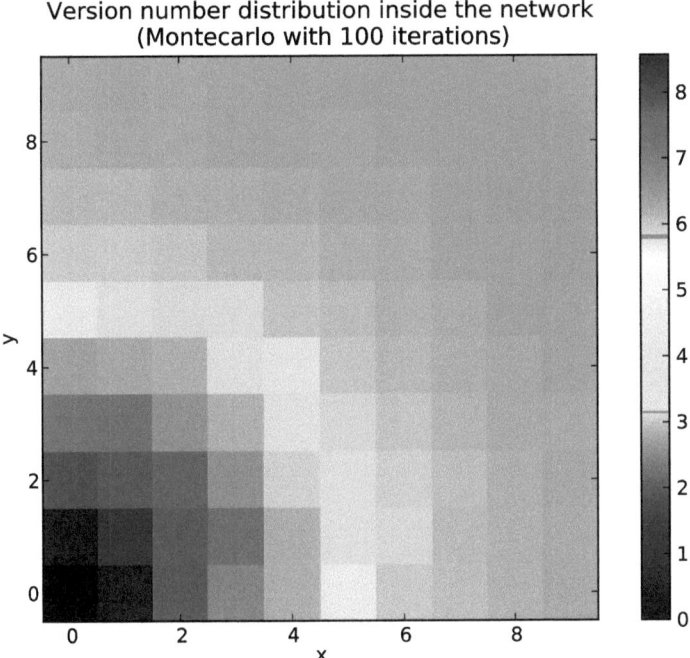

**Figure 6.8:** Distribution of Version Numbers in a 100 Node Grid Scenario [Source: [Ded12]]

| Scenario | | Inter-node Distance | | | | | | | | | | | | | |
|---|---|---|---|---|---|---|---|---|---|---|---|---|---|---|---|
| n | K | 10 | 20 | 30 | 40 | 50 | 60 | 70 | 80 | 90 | 100 | 110 | 120 | 130 | 140 |
| 4 | 1 | 1.88 | 1.92 | 1.96 | 1.92 | 3.08 | 3.00 | 3.52 | 4.68 | 4.92 | 5.04 | 5.00 | 4.88 | 5.28 | >30 |
|  | 3 | 1.92 | 1.88 | 1.92 | 1.92 | 2.96 | 3.16 | 3.44 | 4.84 | 5.16 | 4.84 | 4.72 | 5.00 | 6.64 | 18.96 |
|  | 9 | 1.80 | 1.92 | 1.96 | 1.96 | 3.20 | 3.16 | 4.48 | 5.00 | 4.88 | 4.88 | 4.68 | 4.96 | 4.88 | >30 |
| 9 | 1 | 1.96 | 2.92 | 4.40 | 5.28 | 7.20 | 7.56 | 11.84 | 12.16 | 12.16 | 12.20 | 12.32 | 12.40 | 15.64 | >30 |
|  | 3 | 1.92 | 2.80 | 3.08 | 4.48 | 5.88 | 6.16 | 7.84 | 11.96 | 12.16 | 12.28 | 12.00 | 12.04 | 15.76 | >30 |
|  | 9 | 1.92 | 2.92 | 4.28 | 5.36 | 7.52 | 7.48 | 11.88 | 12.44 | 12.12 | 12.16 | 12.08 | 12.48 | 14.52 | >30 |
| 16 | 1 | 2.56 | 4.52 | 6.80 | 9.04 | 13.48 | 13.40 | 19.88 | 22.48 | 22.00 | 22.32 | 22.20 | 22.52 | 25.64 | >30 |
|  | 3 | 2.36 | 4.08 | 5.72 | 7.60 | 10.48 | 11.56 | 13.80 | 22.48 | 22.04 | 22.24 | 22.36 | 22.08 | 26.32 | >30 |
|  | 9 | 2.32 | 5.28 | 7.08 | 8.84 | 12.80 | 12.72 | 19.24 | 22.60 | 22.12 | 22.04 | 22.92 | 22.48 | 29.16 | >30 |
| 25 | 1 | 4.00 | 6.56 | 10.88 | 14.32 | 19.84 | 19.64 | >30 | >30 | >30 | >30 | >30 | >30 | >30 | >30 |
|  | 3 | 3.04 | 5.64 | 8.72 | 11.80 | 16.36 | 17.80 | 21.32 | >30 | >30 | >30 | >30 | >30 | >30 | >30 |
|  | 9 | 3.88 | 6.88 | 10.52 | 13.92 | 20.96 | 21.20 | >30 | >30 | >30 | >30 | >30 | >30 | >30 | >30 |
| 36 | 1 | 5.00 | 10.28 | 15.80 | 19.48 | 29.20 | 29.56 | >30 | >30 | >30 | >30 | >30 | >30 | >30 | >30 |
|  | 3 | 3.96 | 7.76 | 12.44 | 16.60 | 23.44 | 25.28 | >30 | >30 | >30 | >30 | >30 | >30 | >30 | >30 |
|  | 9 | 5.36 | 9.00 | 15.12 | 21.56 | >30 | 29.92 | >30 | >30 | >30 | >30 | >30 | >30 | >30 | >30 |
| 49 | 1 | 6.32 | 14.04 | 19.28 | 29.24 | >30 | >30 | >30 | >30 | >30 | >30 | >30 | >30 | >30 | >30 |
|  | 3 | 5.24 | 10.24 | 16.40 | 22.96 | >30 | >30 | >30 | >30 | >30 | >30 | >30 | >30 | >30 | >30 |
|  | 9 | 6.56 | 14.20 | 20.28 | >30 | >30 | >30 | >30 | >30 | >30 | >30 | >30 | >30 | >30 | >30 |
| 64 | 1 | 8.80 | 16.00 | 28.20 | >30 | >30 | >30 | >30 | >30 | >30 | >30 | >30 | >30 | >30 | >30 |
|  | 3 | 7.04 | 13.28 | 21.12 | 29.36 | >30 | >30 | >30 | >30 | >30 | >30 | >30 | >30 | >30 | >30 |
|  | 9 | 7.80 | 16.12 | 27.00 | >30 | >30 | >30 | >30 | >30 | >30 | >30 | >30 | >30 | >30 | >30 |
| 81 | 1 | 10.16 | 20.28 | >30 | >30 | >30 | >30 | >30 | >30 | >30 | >30 | >30 | >30 | >30 | >30 |
|  | 3 | 7.96 | 16.60 | 28.72 | >30 | >30 | >30 | >30 | >30 | >30 | >30 | >30 | >30 | >30 | >30 |
|  | 9 | 10.32 | 21.76 | >30 | >30 | >30 | >30 | >30 | >30 | >30 | >30 | >30 | >30 | >30 | >30 |
| 100 | 1 | 11.88 | 26.00 | >30 | >30 | >30 | >30 | >30 | >30 | >30 | >30 | >30 | >30 | >30 | >30 |
|  | 3 | 10.16 | 20.84 | >30 | >30 | >30 | >30 | >30 | >30 | >30 | >30 | >30 | >30 | >30 | >30 |
|  | 9 | 11.76 | 25.04 | >30 | >30 | >30 | >30 | >30 | >30 | >30 | >30 | >30 | >30 | >30 | >30 |
| 225 | 1 | 27.96 | >30 | >30 | >30 | >30 | >30 | >30 | >30 | >30 | >30 | >30 | >30 | >30 | >30 |
|  | 3 | 22.28 | >30 | >30 | >30 | >30 | >30 | >30 | >30 | >30 | >30 | >30 | >30 | >30 | >30 |
|  | 9 | 25.48 | >30 | >30 | >30 | >30 | >30 | >30 | >30 | >30 | >30 | >30 | >30 | >30 | >30 |

**Table 6.2: Simulated 95th Percentiles of the Delay for the Line Scenarios in [s]**

## 6.3 Simulation Results

| Scenario | | | | | | | Inter-node Distance | | | | | | | |
|---|---|---|---|---|---|---|---|---|---|---|---|---|---|---|
| n | K | 10 | 20 | 30 | 40 | 50 | 60 | 70 | 80 | 90 | 100 | 110 | 120 | 130 | 140 |
| 4 | 1 | 1.92 | 1.96 | 1.88 | 1.92 | 1.96 | 1.88 | 1.92 | 1.92 | 1.96 | 2.72 | 3.28 | 3.20 | 3.20 | 7.64 |
|   | 3 | 1.84 | 1.96 | 1.84 | 1.96 | 1.96 | 1.96 | 1.96 | 1.84 | 1.92 | 2.76 | 3.20 | 3.12 | 3.16 | 7.32 |
|   | 9 | 1.92 | 1.96 | 1.92 | 1.92 | 1.96 | 1.84 | 1.96 | 1.92 | 1.92 | 2.72 | 3.12 | 3.12 | 3.20 | 8.44 |
| 9 | 1 | 1.96 | 1.84 | 1.92 | 1.92 | 1.96 | 2.76 | 3.56 | 7.00 | 6.92 | 6.12 | 5.68 | 5.88 | 6.64 | 16.20 |
|   | 3 | 1.92 | 1.88 | 1.92 | 1.96 | 1.96 | 2.84 | 3.24 | 3.40 | 3.48 | 4.20 | 5.60 | 5.44 | 5.96 | 10.04 |
|   | 9 | 1.92 | 1.96 | 1.96 | 1.96 | 1.96 | 2.64 | 3.88 | 7.16 | 7.08 | 6.84 | 5.80 | 5.92 | 6.48 | 11.72 |
| 16 | 1 | 1.96 | 1.96 | 1.88 | 2.76 | 3.76 | 4.60 | 5.56 | 7.68 | 8.24 | 8.40 | >30 | >30 | 20.32 | 17.40 |
|   | 3 | 1.92 | 1.92 | 1.96 | 2.60 | 3.00 | 3.28 | 4.44 | 4.92 | 4.68 | 5.88 | 7.84 | 7.92 | 8.08 | 11.28 |
|   | 9 | 1.92 | 1.96 | 1.96 | 2.80 | 4.24 | 4.32 | 5.36 | 8.36 | 8.08 | 7.76 | >30 | >30 | >30 | 22.72 |
| 25 | 1 | 1.88 | 1.92 | 2.56 | 4.12 | 5.52 | 5.68 | 6.92 | 9.32 | 9.80 | 10.24 | 12.00 | 11.56 | 12.24 | 24.92 |
|   | 3 | 1.96 | 1.92 | 2.52 | 2.96 | 4.04 | 4.24 | 5.36 | 6.12 | 6.32 | 7.52 | 10.12 | 10.28 | 10.36 | 14.60 |
|   | 9 | 1.96 | 1.92 | 2.60 | 3.92 | 5.80 | 6.84 | 7.32 | 9.60 | 9.76 | 10.00 | 11.44 | 11.64 | 12.88 | 22.32 |
| 36 | 1 | 1.96 | 1.96 | 2.92 | 5.08 | 5.80 | 6.40 | 8.08 | 10.68 | 10.36 | 12.28 | 16.04 | >30 | 15.12 | 26.80 |
|   | 3 | 1.92 | 1.92 | 3.04 | 3.56 | 4.40 | 4.96 | 6.64 | 7.64 | 7.68 | 8.68 | 12.16 | 12.16 | 12.44 | 18.60 |
|   | 9 | 1.96 | 1.96 | 3.16 | 4.20 | 6.28 | 6.28 | 7.76 | 10.84 | 11.40 | 11.72 | 24.40 | 20.88 | 14.96 | 22.16 |
| 49 | 1 | 1.92 | 2.60 | 4.08 | 5.24 | 6.84 | 7.76 | 9.08 | 11.92 | 12.76 | 13.84 | 24.20 | >30 | 20.20 | 24.28 |
|   | 3 | 1.96 | 2.44 | 3.20 | 4.04 | 5.24 | 5.72 | 7.72 | 8.84 | 8.80 | 10.08 | 14.08 | 14.52 | 14.52 | 19.52 |
|   | 9 | 1.92 | 2.48 | 4.08 | 4.96 | 6.76 | 7.44 | 9.12 | 12.08 | 12.04 | 14.40 | >30 | >30 | 19.56 | >30 |
| 64 | 1 | 1.96 | 2.84 | 4.56 | 5.68 | 7.36 | 8.16 | 10.92 | 12.92 | 14.36 | 14.76 | >30 | >30 | 21.84 | >30 |
|   | 3 | 1.96 | 2.84 | 3.76 | 4.44 | 5.76 | 6.36 | 8.84 | 10.16 | 10.24 | 11.44 | 16.36 | 16.88 | 17.68 | 22.40 |
|   | 9 | 1.92 | 2.88 | 4.36 | 5.52 | 7.40 | 7.72 | 10.80 | 13.84 | 13.88 | 14.72 | >30 | >30 | 24.56 | >30 |
| 81 | 1 | 1.96 | 3.44 | 5.76 | 6.00 | 8.20 | 8.92 | 11.48 | 14.64 | 14.56 | 16.52 | 23.88 | 23.48 | 24.92 | >30 |
|   | 3 | 1.96 | 2.80 | 4.04 | 5.04 | 6.68 | 7.12 | 10.32 | 11.32 | 11.44 | 13.48 | 18.28 | 19.76 | 19.80 | 25.32 |
|   | 9 | 1.92 | 3.12 | 4.68 | 6.08 | 8.24 | 8.80 | 13.16 | 15.16 | 15.68 | 16.44 | 22.88 | 24.00 | 23.32 | >30 |
| 100 | 1 | 1.96 | 4.04 | 4.88 | 6.64 | 8.76 | 9.80 | 12.92 | 16.88 | 16.60 | 18.32 | 27.80 | 27.08 | 26.28 | >30 |
|   | 3 | 1.88 | 4.76 | 4.20 | 5.56 | 7.20 | 8.00 | 10.84 | 12.96 | 12.60 | 14.28 | 20.60 | 21.16 | 22.36 | 29.40 |
|   | 9 | 1.96 | 3.48 | 5.28 | 6.56 | 9.04 | 9.80 | 13.08 | 16.68 | 16.48 | 19.76 | 25.84 | >30 | 25.92 | >30 |
| 225 | 1 | 3.04 | 5.56 | 7.28 | 10.12 | 13.20 | 15.76 | 18.52 | 23.28 | 24.88 | 27.04 | >30 | >30 | >30 | >30 |
|   | 3 | 2.84 | 4.16 | >30 | 8.44 | 10.32 | 11.52 | 16.32 | >30 | 19.24 | >30 | >30 | >30 | >30 | >30 |
|   | 9 | 2.72 | 5.04 | 7.04 | 9.92 | 13.08 | 14.48 | 18.76 | 24.28 | 23.76 | 28.48 | >30 | >30 | >30 | >30 |

**Table 6.3: Simulated 95th Percentiles of the Delay for the Grid Scenarios in [s]**

|  | Trickle | Push | | |
| --- | --- | --- | --- | --- |
|  |  | $t_{push} = 3s$ | $t_{push} = 5s$ | $t_{push} = 39s$ |
| $t_{95\%}$ | 29.4 s | 27.9 s | 37.0 s | 169.1 s |
| $t_{max}$ | 34.9 s | 33.0 s | 45.2 s | 240.5 s |
| Packets sent until consistent | 176 | 714 | 196 | 415 |
| Packets sent in steady state | $\approx 2.5\frac{1}{s}$ | $33.4 \frac{1}{s}$ | $20 \frac{1}{s}$ | $2.6 \frac{1}{s}$ |

**Table 6.4:** Comparison of Trickle and Push Algorithm [Source: [Ded12]]

| Parameter | Value |
| --- | --- |
| $\tau_L$ | 256 ms |
| $\tau_H$ | 262.144 s |
| $K$ | 3 |

**Table 6.5:** Trickle Algorithm Parameter Values for RPL

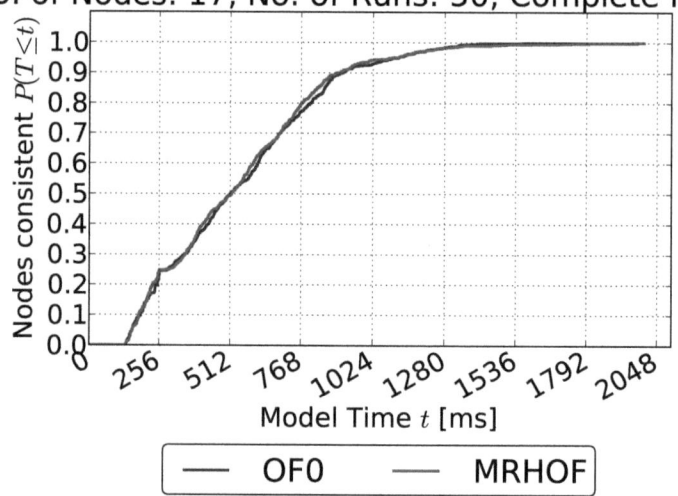

**Figure 6.9:** Time to Discovery of Default Route in the Container Scenario

# 7 Analytical Modelling of Service Discovery

Since simulation runs of the service discovery algorithms for large scenarios and with an appropriate number of Monte Carlo iterations to reach statistical confidence take a considerable amount of time, analytical models have been created. Those models can be calculated in a duration, which is an order of magnitude smaller than the duration for the simulation, cf. section 9.1.3.

## 7.1 Models for the Service Distribution

The main factors governing the efficiency of the Trickle algorithm are the number of messages issued by the algorithm and the delay until the network has reached consistency. Analytical models have been created for both metrics and will be presented in the following sections.

### 7.1.1 Analytical Model for the Time to Consistency

The model for the complete network consistency delay distribution is created from the individual nodes' delay distribution.

The analytical model is based on the fact that the Trickle algorithm draws uniformly distributed pseudo random numbers between $\frac{\tau}{2}$ and $\tau$. If an inconsistency is detected, the algorithm immediately sets $\tau = \tau_L$.

For the simplest scenario 'Line-Direct' as described in section 5.2.1.1, the consistency model can be set up in the following way: The seed of the inconsistency is the 0th hop. It does not draw a random number, but immediately knows the consistent information, thus this results in a Dirac impulse at $t = 0$. That particular node chooses its time to send the new information, which is known to be inconsistent with other information in the network, at a time uniformly distributed in the TX part of the Trickle period between $\frac{\tau_L}{2}$ and $\tau_L$ indicated by the vertical bar, cf. figure 7.1. The 1st hop neighbour (assuming a perfect link for the moment and neglecting processing and communication time) thus detects the inconsistency uniformly distributed between $\frac{\tau_L}{2}$ and $\tau_L$. It will then start its own Trickle period and randomly select a timer in the TX part of the Trickle period and send the information when the timer fires. For the 2nd hop two uniformly distributed random variables are added. The convolution of the two random variables leads to a triangle distribution as shown in figure 7.2. The 3rd hop adds another uniformly distributed random variable resulting in the bell shape shown in the same figure. The central limit theorem states that the mean of a summation of independent and identically distributed random variables, each with finite mean and variance, will be approximately normally distributed. For a larger number of hops, the node consistency distribution will become almost normally distributed.

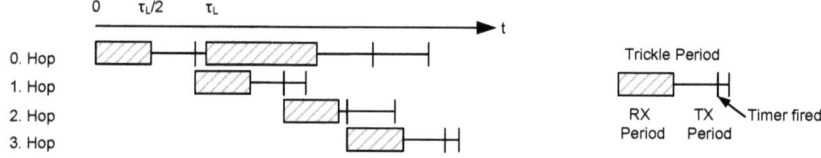

Figure 7.1: Consistency Delay Addition

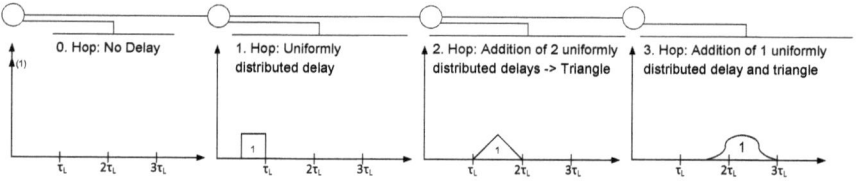

Figure 7.2: Node Consistency Delay Distribution

The Probability Density Function (pdf) of the time to consistency scenario can be modeled in detail by

$$p(t) = \frac{1}{N} \sum_{h=0}^{N-1} \sum_{c=0}^{C'-1} \sum_{a=1}^{N-1} f_{h,c,a}(t) \cdot p_{h,c,a}(t) \quad (7.1)$$

where
$h$: hops
$c$: Trickle cycle
$a$: number of 1-hop ancestors closer to source
$N$: total number of nodes
$C'$: maximum number of Trickle cycles to take into account

$f_{h,c,a}(t)$ are the base distributions which are created by the Trickle algorithm at hop $h$ in cycle $c$ with $a$ ancestors in the previous hop.

$p_{h,c,a}(t)$ describes the relative frequency (occurence) with which the distribution $f_{h,c,a}(t)$ is present.

## 7.1 Models for the Service Distribution

The overall algorithm to calculate cdf $P(T \leq t)$, the base distributions $f_{h,c,a}(t)$ and the occurences $p_{h,c,a}(t)$ are shown in algorithms 7.1, 7.2, and 7.3.

The base distributions $f_{h,c,a}(t)$ are calculated and plotted in algorithm 7.1. The details on the calculation are presented in section 7.1.1.1.

```
calc_base_distributions()
plot_base_distributions(distributions)
```

**Algorithm 7.1:** Calculation of $f_{h,c,a}(t)$

```
calc_prr(scenario)
calc_neighbours(prr)
calc_tx_outcomes(neighbours, prr)
calc_hopcounts(prr, inject_node, trickle_k, neighbours, scenario)
calc_hopcounts_next_cycles(hopcounts)
calc_neighbours_hop_closer(hopcounts, prr, trickle_k)
calc_prob_mix_from_hopcount(hopcounts)
```

**Algorithm 7.2:** Calculation of $p_{h,c,a}(t)$

In order to derive $p_{h,c,a}(t)$, the PRR matrix is calculated based on the scenario description. From the PRR matrix the neighbours of all nodes, can be deducted as well as the possible transmission outcomes of each node and its associated probabilities. Using this information the probabilities for a certain hopcount can be calculated. This result can be used to calculate the amount of next Trickle cycle transmissions and the average number of ancestors at each hopcount. The probabilities for each node are then used to derive $p_{h,c,a}(t)$. More details are found in section 7.1.1.2.

Finally, from $f_{h,c,a}(t)$ and $p_{h,c,a}(t)$ the cdf $P(T \leq t)$ can be derived according to equation 7.1 and plotted as shown in algorithm 7.3.

```
calc_cdf(prob_mix, distributions)
plot_graphs(cdf)
```

**Algorithm 7.3:** Calculation of $P(T \leq t)$

### 7.1.1.1 Base Distributions $f_{h,c,a}(t)$

$f_{h,c,a}(t)$ can be calculated according to equation 7.2.

$$f_{h,c,a}(t) = \begin{cases} \delta(t) & , h = 0, \\ \mathcal{L}^{-1}\{\mathcal{L}\{\Theta(t-\frac{T_L}{2}) - \Theta(t-\tau_L)\}^h\} & , h \geq 1, c = 0, a = 1 \\ \mathcal{L}^{-1}\{\mathcal{L}\{\Theta(t-\frac{T_L}{2}) - \Theta(t-\tau_L)\} \cdot \\ \quad \mathcal{L}\{\Theta(t-2\tau_L) - \Theta(t-3\tau_L)\}^{h-1}\} & , h > 0, c = 1, a = 1. \end{cases}$$
(7.2)

$\Theta(\cdot)$ denotes the Heaviside step function. $\mathcal{L}$ denotes the Laplace transform and $\mathcal{L}^{-1}$ denotes the inverse Laplace transform.

The distibutions for more than 1 ancestor ($a > 1$) are based on the $a = 1$ distributions. If there is more than 1 ancestor, the receiving node is informed by the ancestor that is transmitting first, thus the minimum of $a$ Random Variables (RVs) (distributed as decribed above) needs to be used. The cdf of the minimum of $a$ independent and identically distributed variables can be calculated by $1 - (1 - P(X \leq x))^a$ as shown in appendix B.

For the scenario shown in figure 7.2, where the PRR is 1 for the shown paths and PRR is 0 for paths that are not shown, the distributions are as follows:
$f_{0,0,1}(t) = \delta(t)$
$f_{1,0,1}(t) = \mathcal{L}^{-1}\{\mathcal{L}\{\Theta(t-\frac{T_L}{2}) - \Theta(t-\tau_L)\}^1\}$
$f_{2,0,1}(t) = \mathcal{L}^{-1}\{\mathcal{L}\{\Theta(t-\frac{T_L}{2}) - \Theta(t-\tau_L)\}^2\}$
$f_{3,0,1}(t) = \mathcal{L}^{-1}\{\mathcal{L}\{\Theta(t-\frac{T_L}{2}) - \Theta(t-\tau_L)\}^3\}$
$f_{h,\forall c \geq 1, \forall a \geq 2}(t) = 0$ as all information is received in the first Trickle cycle for all nodes.

The frequencies are equally weighting the distributions: $p_{0,0,1}(t) = 1$, $p_{1,0,1}(t) = 1$, $p_{2,0,1}(t) = 1$, and $p_{3,0,1}(t) = 1$. For the particular scenario, drawings of the pdf and cdf are depicted in figure 7.3.

In figure 7.4 the numerically solved pdfs for $p_{h,c,a}(t)$ with $c$ varied and $h = 1, a = 0$ are shown. The typical Trickle behaviour of binary exponentially increasing backoff times is clearly visible. Cycle 3 is partly and cycle 4 is completely outside of the shown time frame.

Numerically solved pdfs for $p_{h,c,a}(t)$ with $h$ varied and $c = 0, a = 0$ are shown in figure 7.5. The Dirac impulse for $h = 0$, the rectangle distribution for $h = 1$ and the increasingly normally distributed distributions for higher $h$ values are clearly visible.

The influence of multiple ancestors on the distribution are shown in figure 7.6 (for various h values). With higher $a$ values (more ancestors) the distributions tend

## 7.1 Models for the Service Distribution

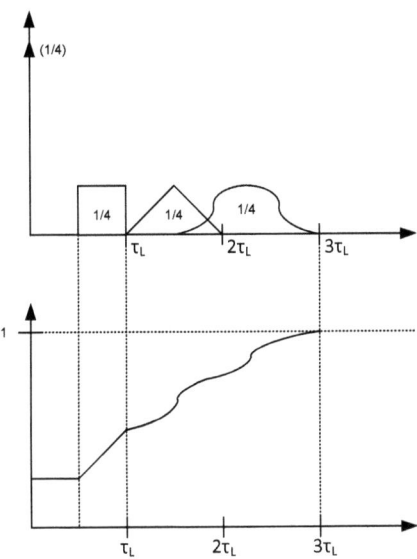

**Figure 7.3: Probability Density and Cumulative Distribution Function of the Network Consistency Delay Distribution**

to have a lower mean and variance. Figure 7.7 shows the same data in a three dimensional presentation split up into subfigures by the value of $a$. Especially for $h = 1$ the change of the distribution shape can be seen between for the different $a$ values. When comparing figures 7.7a to 7.7c it can also be seen that the width of the distribution is reduced and the mean is shifted to lower values.

The influence of higher Trickle cycle $c$ values at higher hopcounts $h$ on the distributions can be seen in figures 7.8 (for $h = 2$) and 7.9 (for $h = 3$).

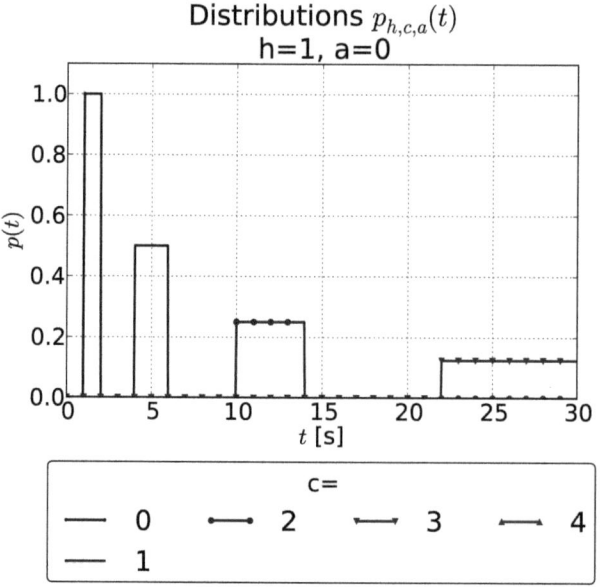

**Figure 7.4:** Probability Density Function for $h = 1, a = 0$, $c$ **varied**

## 7.1 Models for the Service Distribution

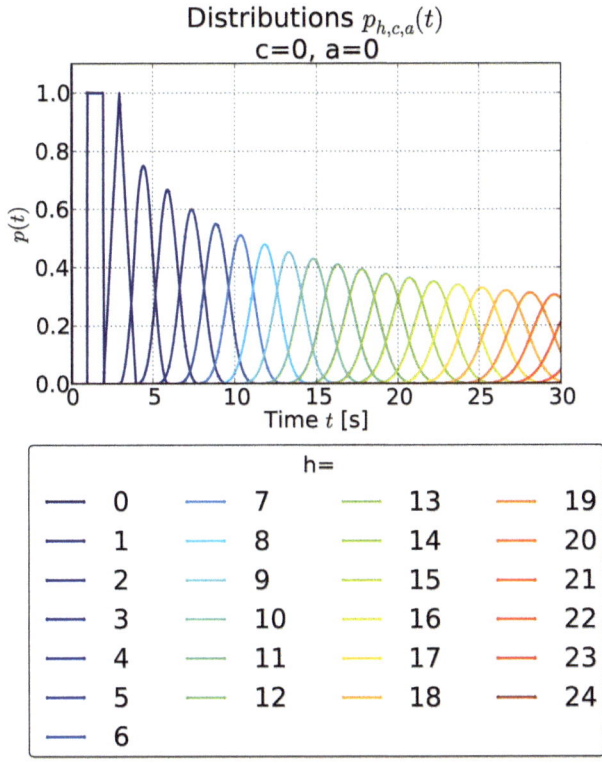

**Figure 7.5:** Probability Density Function for $c = 0, a = 0$, $h$ **varied**

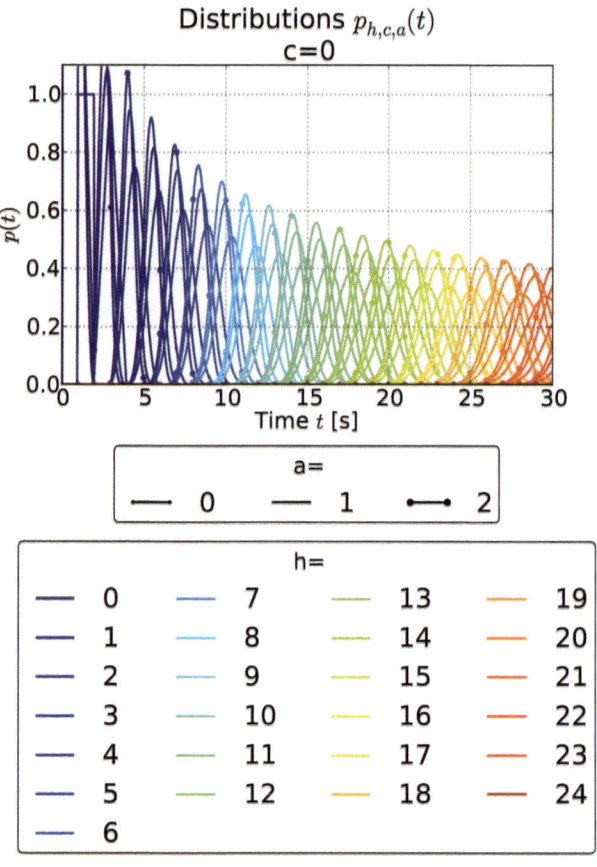

**Figure 7.6:** Probability Density Function for $c = 0$, $h$ and $a$ varied

## 7.1 Models for the Service Distribution

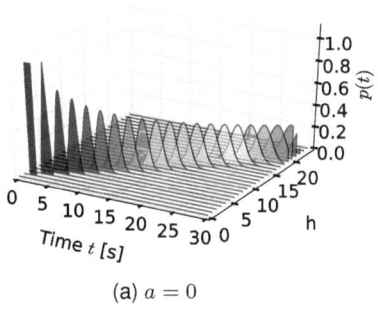

(a) $a = 0$

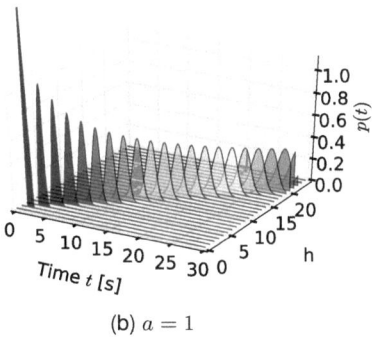

(b) $a = 1$

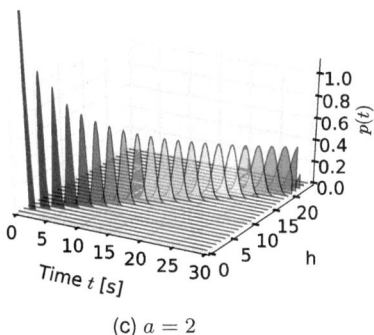

(c) $a = 2$

**Figure 7.7: Probability Density Function for $c = 0$, $h$ and $a$ varied**

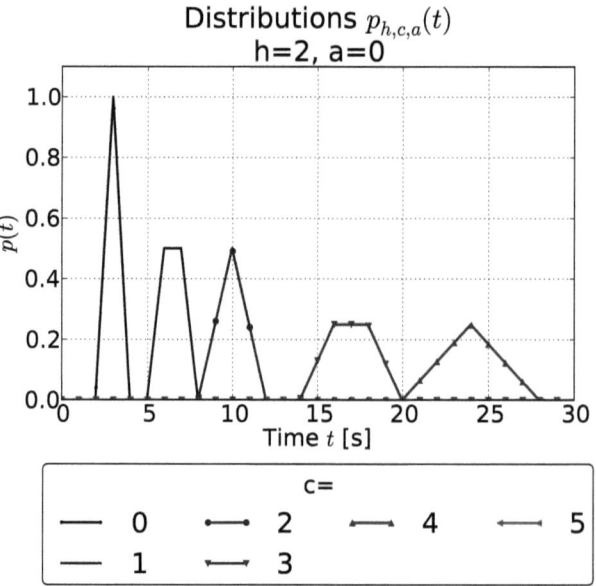

**Figure 7.8:** Probability Density Function for $h = 2, a = 0$, $c$ varied

## 7.1 Models for the Service Distribution

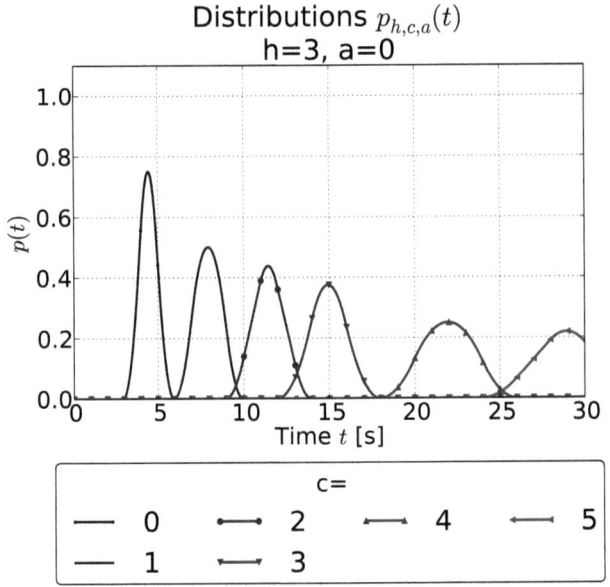

**Figure 7.9:** Probability Density Function for $h = 3, a = 0$, $c$ **varied**

### 7.1.1.2 Relative Frequency $p_{h,c,a}(t)$

The calculation of $p_{h,c,a}(t)$ depends on the scenario type (Line-Direct, Line-CPM, Grid, etc.) and the scenario parameters (number of nodes and inter-node distance).

The algorithm 7.2 starts with the CPM propagation model shown in figure 7.10 and the scenario type and parameters to calculate the PRR matrix with the PRR values between all pairs of nodes. The entries in the PRR matrix are based on the distance between the transmitter $n_{TX}$ and the receiver $n_{RX}$. That distance evaluates to a certain PRR value according to the CPM model (more details are given in appendix F).

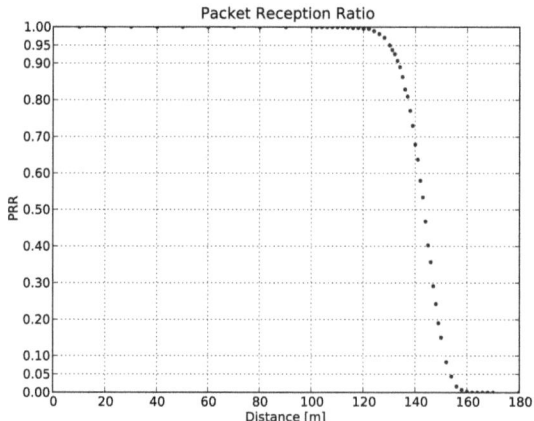

**Figure 7.10: Packet Reception Ratio in a 2 node scenario with CPM propagation model**

E.g. for a sample 4 node Line-CPM scenario as in figure 7.11 the distance between directly neighbouring nodes is determined by the distance between the nodes and results in a certain PRR1 according to the CPM model. The dark round and rectangle nodes are separated by double the base distance, which results in a different (and lower) PRR2. Similarly, the white round and the dark rectangle nodes are separated by a triple base distance which results in PRR3.

For a 9 node Line-CPM 30 m scenario the resulting PRR matrix is listed in table 7.1.

The PRR matrix can then be used to deduct the neighbours for each node and the possible transmission outcome when a certain node transmits (latter is shown in table 7.2). For example, if node 1 transmits in the 9 node Line-CPM 30 m scenario,

## 7.1 Models for the Service Distribution

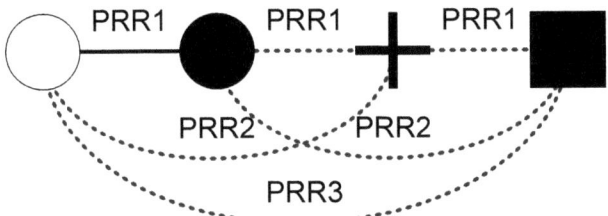

**Figure 7.11:** 4 Node Scenario and Packet Reception Ratios between the Nodes

there are 4 possible transmission outcomes. In the first possible outcome, the nodes 2, 3, 4, 5, 6 will receive the information. This outcome has a probability of ≈ 0.1495. The more likely transmission outcome (with a probability of ≈ 0.8461) is that only nodes 2, 3, 4, 5 will receive and node 6 will not receive. The other two outcomes of node 1 transmitting have low probabilities.

The hopcount probability matrix is calculated by looping over all hopcounts and all nodes. The nodes of the previous hop, that are closer to the service injecting node are searched. For those nodes, the possible transmission sets according to the Trickle parameters $k$ are deduced. Using these and the transmission outcome probabilities which have already been calculated, the probability to be at the current hopcount is calculated.

The result of the hopcount calculation are the probabilities of each node of the scenario to be at a certain hopcount if the node is informed in the first possible Trickle cycle ($c = 0$). An example is shown in table 7.3.

Using the hopcount information of Trickle cycle 0, the probabilities for the next cycles $c \geq 1$ can be calculated using the remaining probabilities.

The average number of ancestors $a$ at each hopcount can be calculated from the hopcount probabilities as well.

Finally the relative frequency $p_{h,c,a}(t)$ can be calculated from the hopcount probabilities.

Graphs showing the results of the analytical model presented here are contained in section 9.1 for comparison of the analytical model and the simulation results.

### 7.1.2 Analytical Model for the Number of Packets Sent

The number of sent packets depends mostly on the number of nodes in the scenario, the distance between the nodes, the number of neighbours of the nodes and the number of Trickle cycles.

The following equation can be used to calculate the number of Trickle cycles $Y$ in the observation period $T_m$:

$$Y = \begin{cases} \log_2 \tau_H - \log_2 \tau_L + 1 + \frac{T_m - (2\tau_H - \tau_L)}{\tau_H} \\ \qquad \qquad \qquad \qquad \text{if } T_m > 2\tau_H - \tau_L, \\ \lfloor \log_2(T_m + \tau_L) \rfloor - \log_2 \tau_L \rfloor + \\ \frac{T_m - 2^{\log_2(\tau_L) + \lfloor \log_2(T_m + \tau_L) - \log_2 \tau_L \rfloor + 2 \lfloor \log_2(T_m + \tau_L) - \log_2 \tau_L \rfloor}}{2^{\log_2(\tau_L) + \lfloor \log_2(T_m + \tau_L) - \log_2 \tau_L \rfloor}} \\ \qquad \qquad \qquad \qquad \text{if } T_m \leq 2\tau_H - \tau_L. \end{cases} \quad (7.3)$$

The first case of the equation denotes the case when the observation period is long enough so that the Trickle algorithm has reached the highest Trickle interval $\tau_H$; while the second case deals with the ramp-up phase of the Trickle algorithm.

The non-fractional part of the equation for the first case denotes the number of Trickle cycles in the Trickle ramp-up period. The enumerator of the fractional part of the equation denotes the remaining $\tau_H$ long cycles.

The non-fractional part of the equation for the second case denotes the number of non-fractional cycles in the Trickle ramp-up period. The enumerator of the fractional part of the equation denotes the left-over length of the last ramp-up period. The denominator gives the full length of the last ramp-up period.

Using equation 7.3 the number of cycles $Y$ can be used to derive the mean number of packets which are sent in the observation period $T_m$. The analytical model for the mean number of packets $\overline{M}$ depends on the Trickle redundancy constant $K$, the number of nodes in the scenario $N$ and the mean number of neighbours $\overline{N^n} = \frac{1}{N} \sum_{i=0}^{N-1} N_i^n$, where $N_i^n$ is the number of neighbours of node $i$.

$$\overline{M} = N * Y * \frac{\min(\overline{N^n} + 1, K)}{\overline{N^n} + 1} \quad (7.4)$$

The fractional part of the equation gives the probability of a node sending in one Trickle cycle. This probability is restricted by the redundancy constant $K$, when the mean number of neighbours is higher. The mean number of neighbours has been derived from the scenario in algorithm 7.2 already. Multiplying the probability with the number of nodes in the scenario and the number of cycles gives the number of sent packets. Results from the analytical model are presented and compared to simulation results in section 9.1.2.

## 7.1 Models for the Service Distribution

| $n_{TX}$ \ $n_{RX}$ | 1 | 2 | 3 | 4 | 5 | 6 | 7 | 8 | 9 |
|---|---|---|---|---|---|---|---|---|---|
| 1 | 1 | 1 | 1 | 1 | 0.995 | 0.15 | 0 | 0 | 0 |
| 2 | 1 | 1 | 1 | 1 | 1 | 0.995 | 0.15 | 0 | 0 |
| 3 | 1 | 1 | 1 | 1 | 1 | 1 | 0.995 | 0.15 | 0 |
| 4 | 1 | 1 | 1 | 1 | 1 | 1 | 1 | 0.995 | 0.15 |
| 5 | 0.995 | 1 | 1 | 1 | 1 | 1 | 1 | 1 | 0.995 |
| 6 | 0.15 | 0.995 | 1 | 1 | 1 | 1 | 1 | 1 | 1 |
| 7 | 0 | 0.15 | 0.995 | 1 | 1 | 1 | 1 | 1 | 1 |
| 8 | 0 | 0 | 0.15 | 0.995 | 1 | 1 | 1 | 1 | 1 |
| 9 | 0 | 0 | 0 | 0.15 | 0.995 | 1 | 1 | 1 | 1 |

**Table 7.1: PRR Matrix for a N=9, d=30m CPM Line Scenario**

| $n_{TX}$ | $n_{RX}$ | $n_{\neg RX}$ | P |
|---|---|---|---|
| 1 | (2, 3, 4, 5, 6) | ∅ | 0.1495.. |
|   | (2, 3, 4, 5) | (6) | 0.8461.. |
|   | (2, 3, 4, 6) | (5) | 0.0007.. |
|   | (2, 3, 4) | (5, 6) | 0.0037.. |
| 2 | (1, 3, 4, 5, 6, 7) | ∅ | 0.1495.. |
|   | (1, 3, 4, 5, 6) | (7) | 0.8461.. |
|   | (1, 3, 4, 5, 7) | (6) | 0.0007.. |
|   | (1, 3, 4, 5) | (6, 7) | 0.0037.. |
| 3 | (1, 2, 4, 5, 6, 7, 8) | ∅ | 0.1495.. |
|   | (1, 2, 4, 5, 6, 7) | (8) | 0.8461.. |
|   | (1, 2, 4, 5, 6, 8) | (7) | 0.0007.. |
|   | (1, 2, 4, 5, 6) | (7, 8) | 0.0037.. |
| 4 | (1, 2, 3, 5, 6, 7, 8, 9) | ∅ | 0.1495.. |
|   | (1, 2, 3, 5, 6, 7, 8) | (9) | 0.8461.. |
|   | (1, 2, 3, 5, 6, 7, 9) | (8) | 0.0007.. |
|   | (1, 2, 3, 5, 6, 7) | (8, 9) | 0.0037.. |
| 5 | (1, 2, 3, 4, 6, 7, 8, 9) | ∅ | 0.9912.. |
|   | (1, 2, 3, 4, 6, 7, 8) | (9) | 0.0044.. |
|   | (2, 3, 4, 6, 7, 8, 9) | (1) | 0.0044.. |
|   | (2, 3, 4, 6, 7, 8) | (1, 9) | 0.0000.. |
| 6 | (1, 2, 3, 4, 5, 7, 8, 9) | ∅ | 0.1495.. |
|   | (2, 3, 4, 5, 7, 8, 9) | (1) | 0.8461.. |
|   | (1, 3, 4, 5, 7, 8, 9) | (2) | 0.0007.. |
|   | (3, 4, 5, 7, 8, 9) | (1, 2) | 0.0037.. |
| 7 | (2, 3, 4, 5, 6, 8, 9) | ∅ | 0.1495.. |
|   | (3, 4, 5, 6, 8, 9) | (2) | 0.8461.. |
|   | (2, 4, 5, 6, 8, 9) | (3) | 0.0007.. |
|   | (4, 5, 6, 8, 9) | (2, 3) | 0.0037.. |
| 8 | (3, 4, 5, 6, 7, 9) | ∅ | 0.1495.. |
|   | (4, 5, 6, 7, 9) | (3) | 0.8461.. |
|   | (3, 5, 6, 7, 9) | (4) | 0.0007.. |
|   | (5, 6, 7, 9) | (3, 4) | 0.0037.. |
| 9 | (4, 5, 6, 7, 8) | ∅ | 0.1495.. |
|   | (5, 6, 7, 8) | (4) | 0.8461.. |
|   | (4, 6, 7, 8) | (5) | 0.0007.. |
|   | (6, 7, 8) | (4, 5) | 0.0037.. |

Table 7.2: TX Outcome for a N=9, d=30m CPM Line Scenario

## 7.1 Models for the Service Distribution

|   |   |   | h |   |
|---|---|---|---|---|
| n | 0 | 1 | 2 | 3 |
| 1 | 1 |   |   |   |
| 2 |   | 1 |   |   |
| 3 |   | 1 |   |   |
| 4 |   | 1 |   |   |
| 5 |   | 0.9956.. | 0.0044.. |   |
| 6 |   | 0.1502.. | 0.8498.. |   |
| 7 |   |   | 1 |   |
| 8 |   |   | 0.9999.. | 0.0000.. |
| 9 |   |   | 0.9937.. | 0.0063.. |

**Table 7.3: Hopcount probabilities for a N=9, d=30m CPM line scenario**

# 8 Measurements of Service Discovery in Wireless Sensor Networks

## 8.1 Measurement Setup

The setup for the measurement of service discovery on a real Wireless Sensor Network is as follows. The sensor nodes in use are XBow/MEMSIC TelosB [PSC05] as shown in figure 8.1.

**Figure 8.1:** Wireless Sensor Node TelosB

A network of 9 nodes arranged in a grid layout at an office room ceiling has been set up, depicted in figure 8.2. The inter-node distance is 120 cm and the nodes are 40 cm beneath the ceiling. The nodes are powered by the USB connection, which is also used for measurement control and logging.

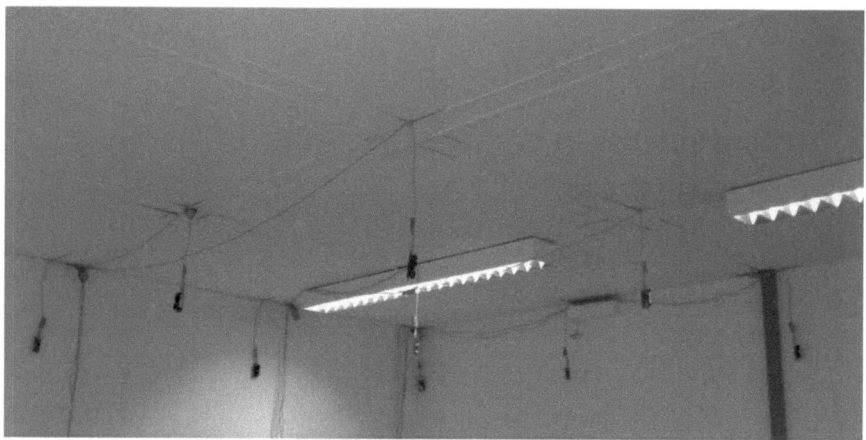

**Figure 8.2: Ceiling Network**

### 8.1.1 Link Assessment Measurements

To vary the connectivity between the nodes, the transmission power level of the nodes is adapted. The setting of the radio chip CC2420 governing the transmission output power is called PA_LEVEL, cf. table 9 of the CC2420 datasheet [Tex07]. In [PP08] the values of the datasheet have been interpolated using a cubic spline interpolation, the resulting dBm values are listed in table 8.1.

As the distance in the testbed is low, low values for the PA_LEVEL have been chosen as well. The resulting PRR topologies have been measured for PA_LEVEL=0 to 3 out of the possible range 0 to 31. For the link measurements the nodes have been programmed with a dedicated application, that listens for received packets and forwards those packets over the USB interface. Additionally, the program can be triggered over the USB interface to transmit a specified number of packets. In turn, each node of the testbed has been triggered to transmit 100 packets with an inter-packet interval of 100 ms. Only one node has been transmitting at a time in order to avoid collisions. After a node has transmitted the 100 packets, the next node is instructed to transmit. This repeats until all nodes have been acting as the transmitter. All nodes not acting as transmitter, are acting as receiver and report the received packets over the USB interface.

Scripts automating the installation, the setup on the host computer, the test execution and the post-processing have been developed. The results of the testbed link evaluation will be presented in the following.

## 8.1 Measurement Setup

| PA_LEVEL | Power (dBm) |
|---|---|
| 31 | 0 |
| 30 | -0.0914 |
| 29 | -0.3008 |
| 28 | -0.6099 |
| 27 | -1.0000 |
| 26 | -1.4526 |
| 25 | -1.9492 |
| 24 | -2.4711 |
| 23 | -3.0000 |
| 22 | -3.5201 |
| 21 | -4.0275 |
| 20 | -4.5212 |
| 19 | -5.0000 |
| 18 | -5.4670 |
| 17 | -5.9408 |
| 16 | -6.4442 |
| 15 | -7.0000 |
| 14 | -7.6277 |
| 13 | -8.3343 |
| 12 | -9.1238 |
| 11 | -10.0000 |
| 10 | -10.9750 |
| 9 | -12.0970 |
| 8 | -13.4200 |
| 7 | -15.0000 |
| 6 | -16.8930 |
| 5 | -19.1530 |
| 4 | -21.8370 |
| 3 | -25.0000 |
| 2 | -28.6970 |
| 1 | -32.9840 |
| 0 | -37.9170 |

**Table 8.1: Power Levels of the CC2420 Radio Chip [Source: [PP08]]**

## 8 Measurements of Service Discovery in WSNs

With a PA_LEVEL=0 as shown in figure 8.3 no full connectivity is possible. Even some links between neighbouring nodes have a low PRR or no connectivity at all. Differences in the PRR of the links can be attributed to manufacturing differences of the radio chip (the datasheet allows variations from 0 to -3dBm for PA_LEVEL=31), differences of the hardware platform (e.g. antenna matching), slightly different alignment of the nodes, different distances to the walls as well as local differences in the noise floor.

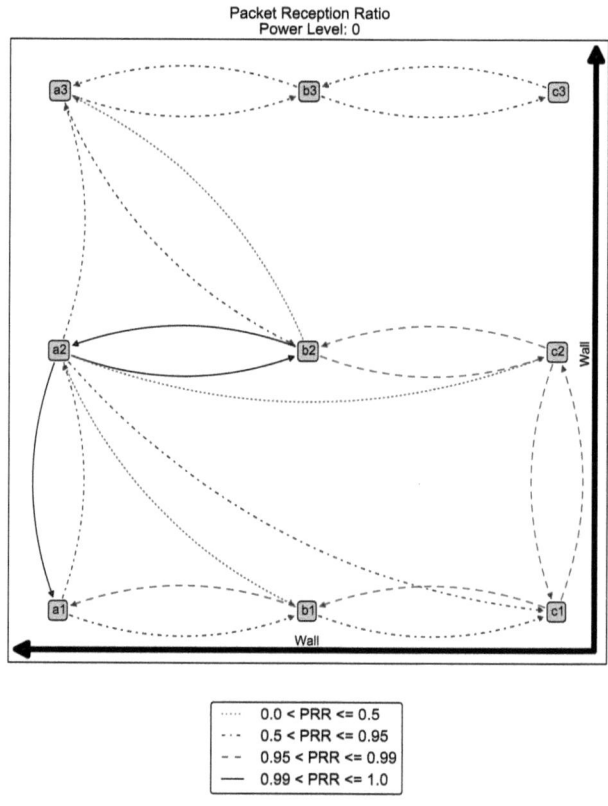

**Figure 8.3: Measured Link-layer Packet Reception Ratios at Transmission Power Setting 0**

## 8.1 Measurement Setup

With a PA_LEVEL=1, the weaker links of the previous setting have been improved to PRRs above 0.95 and additional low PRR links appear, cf. figure 8.4.

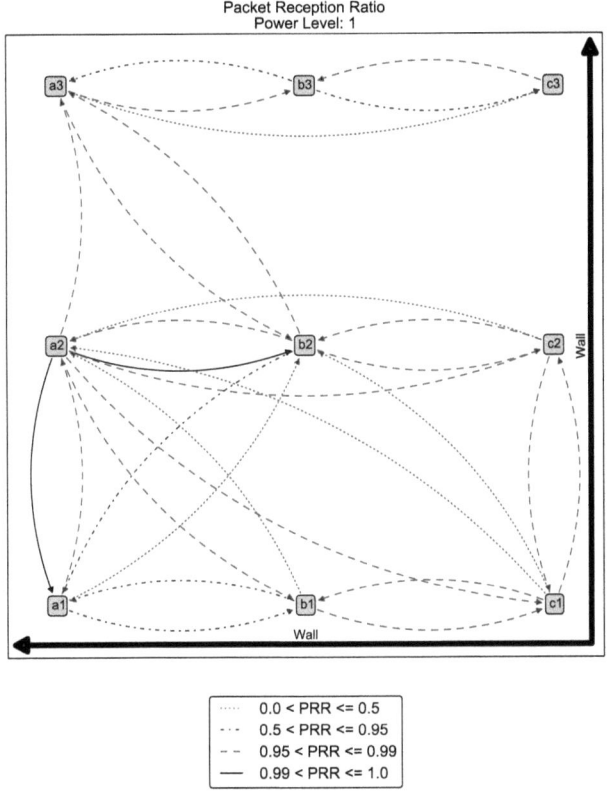

**Figure 8.4: Measured Link-layer Packet Reception Ratios at Transmission Power Setting 1**

With a PA_LEVEL=2, all direct neighbour links are present (with low PRR for some some links though). There are some links between nodes that have the largest distance in the testbed (i.e. nodes a1 and c3), cmp. 8.5. It can be seen that links between direct neighbors (e.g. c2/c3) can even be worse than links that are over longer distances (e.g. c1/c3). This again can be attributed to local interference and noise levels, variations in the hardware, antenna alignment differences, etc.

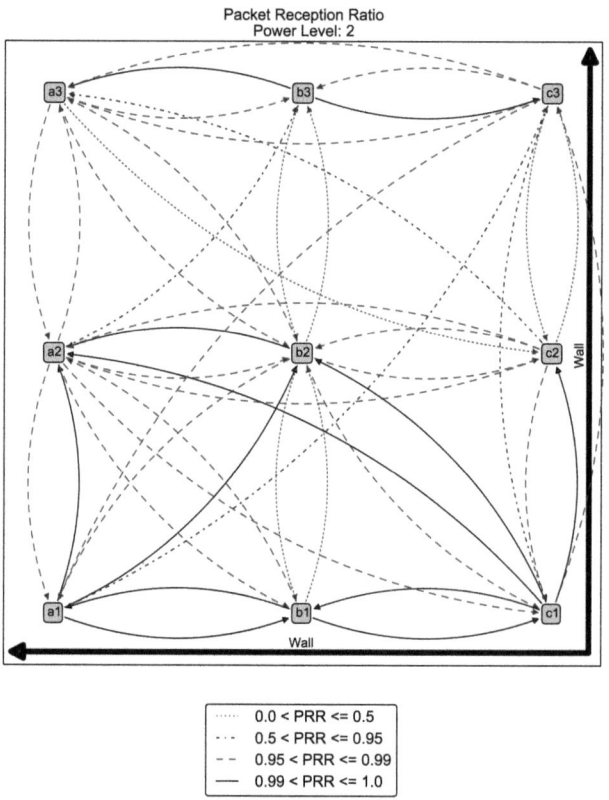

**Figure 8.5: Measured link-layer Packet Reception Ratios at Transmission Power Setting 2**

## 8.1 Measurement Setup

Having set the PA_LEVEL to 3 gives the PRRs as shown in figure 8.6. It can be seen that almost all links between all nodes have reached good PRR values above 0.95.

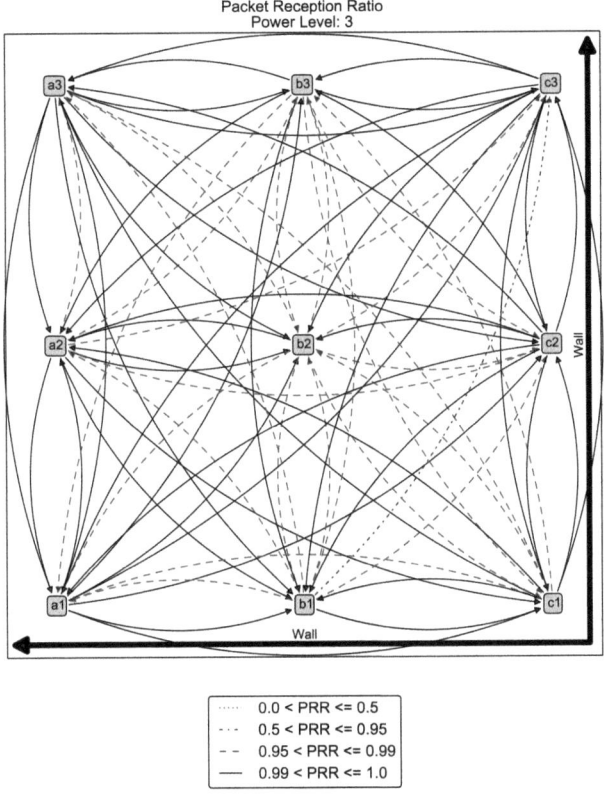

**Figure 8.6: Measured link-layer Packet Reception Ratios at Transmission Power Setting 3**

When comparing the measurement results with the simulation, the exact same link behaviour in a grid scenario cannot be reached due to the above mentioned reasons (manufacturing differences of the radio chip, differences of the hardware platform, slightly different alignment of the nodes, different distances to the walls as well as local differences in the noise floor). Nevertheless, a rough categorisation of the different PA_LEVELs to distances in the simulation scenarios at a PA_LEVEL=0 can be made.

For example, the PA_LEVEL=3 with its almost full connectivity can be compared to a distance of 10m (compare the simulated topologies shown in figure E.1). PA_LEVEL=2 corresponds to a distance range of about 50m (as in figure E.2), while PA_LEVEL=1 corresponds to a distance range of about 110m (compare figure E.3). Finally, PA_LEVEL=0 corresponds to a distance of about 145m (as depicted in figure E.4). Since, these are only approximate equivalences, further simulations have been performed, which use the measured link properties. Note that in most real applications the nodes would be working with PA_LEVEL=31; the reduced PA_LEVELs have been used here to mimic low density topologies in a space limited testbed.

## 8.2 Measurement Results

The same Service Discovery (SD) implementation which has been used to gain the simulation results presented in section 6.2 is also used for measurements. It has been slightly altered to output the debug information and enable the control of the experiment over the USB interface. The measurements have been performed in the testbed with varied PA_LEVELs as described above. Several Monte-Carlo runs have been performed for each parameter setting.

The time of the injection of a service and the time of the discovery of a new service is logged and can be used to calculate the consistency delay as it has been done for the simulation as well. Additionally, all sent packets by each node are logged and can be used to calculate the total number of sent packets.

The measurement results will be shown in comparison to the simulation results in section 9.2.

# 9 Evaluation of Analytical Modelling, Simulation and Measurements

In the next section the results of only the analytical model and the simulation are evaluated. The measurement results will be compared to the results of the model and the simulation in section 9.2.

## 9.1 Comparison of Analytical Model and Simulation Results

The Trickle parameters that have been used for the simulation and the analytical model are listed in table 9.1. The major factor governing the Distribution Delay discussed in section 9.1.1 is the parameter $\tau_L$. The chosen value does not influence the behaviour in any other way as a time-wise scaling. Thus the results shown later can be applied to other values of $\tau_L$ as well. The parameter $\tau_H$ was chosen in such a way that the biggest interval can be reached and kept for some time within the simulation time. For the Trickle parameter $K$ (which determines how often a nodes needs to receive consistent information from other nodes) typically used values in the range that is given in RFC 6206 [Lev+11] have been chosen. A value of 1 for $K$ is chosen as a representative for a low setting, restricting the number of messages send in a neighbourhood to 1, resulting also in the minimal messages to be sent. In a dense neighbourhood, $K = 1$ allows only one node to sent in the complete neighbourhood. For sparse neighbourhoods (e.g. only 1 neighbouring node), the low value will considerably increase the delay, since the neighbour will be forced to not transmit in the current cycle and only with a 50% probability in the next cycle as well. $K = 3$ is chosen as a reasonable trade-off between low delay and low number of sent messages. $K = 9$ is an example of a rather high setting, resulting in low delay and a high number of packets sent.

| Parameter | Value |
|---|---|
| $\tau_L$ | 2 s |
| $\tau_H$ | 32 s |
| $K$ | 1, 3 or 9 |

**Table 9.1: Trickle Algorithm Parameter Value**

In the following, several comparisons are shown for different scenario types, various number of nodes and varying distances. The parameter $a$ (the number of 1-hop ancestors closer to the source) and $h$ (hopcount) of the analytical model are influenced by the scenario type, number of nodes $n$, and the parameter $K$. $a$ increases with the number of nodes $n$ and $K$. In a grid $a$ is also higher because of the higher number of neighbours for each node. $h$ increases with $n$ and is higher for Line scenarios compared to the Grid scenarios.

### 9.1.1 Distribution Delay

In figure 9.1 the combined cdf of the delay of all nodes in a scenario is depicted. The dashed lines show the analytical results, while the solid lines show the simulated results of a 'Line-CPM' scenario with 4 nodes and the Trickle parameter $K$ has been set to 3. Various distances are shown. The other Trickle parameters are listed in table 9.1. $\tau_H$ does not have any influence, as the Trickle algorithm immediately uses $\tau_L$ when an inconsistency is detected. The results shown are for a $\tau_L = 2$ s. The results can however be scaled for other values of $\tau_L$, since the base distributions are parameterised with $\tau_L$.

Since the initial node is instantly informed of the service, 1 out of 4 nodes or 25 % are informed and thus the initial jump at $t = 0$ can be explained.

For inter-node distances of less than 50 m all other 3 out of the 4 nodes are completely in range of the initial node according to the 'Line-CPM' scenario, which leads to the linear increase from 25 % to 100 %, which resembles the uniformly distributed random number that is chosen once. The distances between 80 m and 125 m conform to the typical expected distribution that was expressed in figure 7.3, which is also the typical distribution for the 'Line-Direct' scenario. At those inter-node distances the CPM model results in a PRR of 0 between non-neighbouring nodes and a PRR of 1 between directly neighbouring nodes. Distances between 50 m and 70 m are a mix of the previously two described cases. For inter-node distances above 125 m the CPM propagation model gives PRR values lower than 1 even for directly neighbouring nodes, so that not every transmission is successfully received by the next closest node in the line. This leads to the cdf being dominated by distributions, which are created by convolutions of broader unit functions (for later Trickle cycles) and of the base unit function.

From the figure the close fit between the analytical model and the simulated results can be seen for all distances in a Line-CPM scenario for a value of $K = 3$. For the distance 50 m, where the delay is highly dominated by the next two Trickle cycles transmissions a slight under-estimation can be seen for cdf values of $t > 12s$, thereby giving a worst-case estimation of the delay. Typically a network

## 9.1 Comparison of Analytical Model and Simulation Results

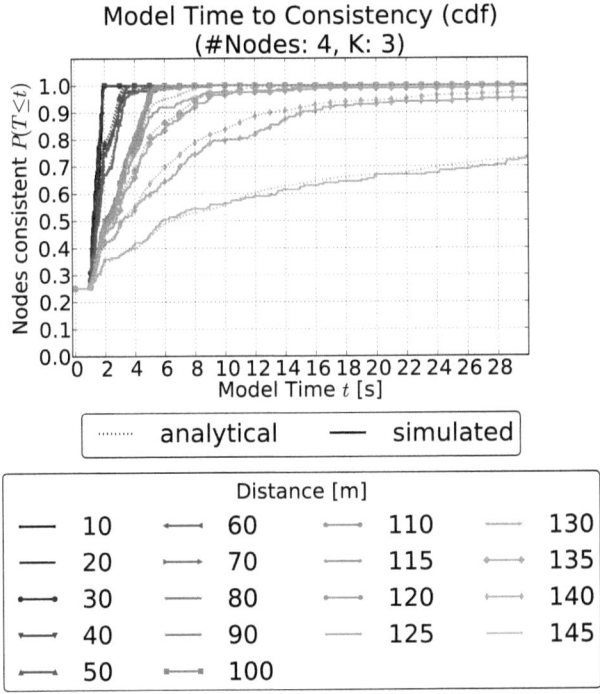

**Figure 9.1:** Network Consistency Delay Distribution
(Line-CPM Scenario, 4 Nodes, K=3)

would not be operated with these inter-node distance which give such low PRR values for all links.

In figure 9.2 the delay distribution for a $K$ value of 1 is shown. The lower $K$ value restricts the number of transmissions in a transmission/collision domain. For the 4 node scenario the major difference to figure 9.1 is for a distance of 70 m between $t = 3s$ and $t = 4s$, which is significantly lower for $K = 1$. This is due to the fact that at lower distances, nodes are likely to be at a closer hop count from the service seeding node. While at higher distances, nodes are so far apart that they are not part of the same transmission/collision domain. Only direct neighbours are then in the same domain and only the preceeding neighbour can have consistent information and thus transmit, those distances are not limited by $K$. For a scenario with 4 nodes, higher values of K are not showing any different behaviour than the behaviour that can be seen at $K = 3$.

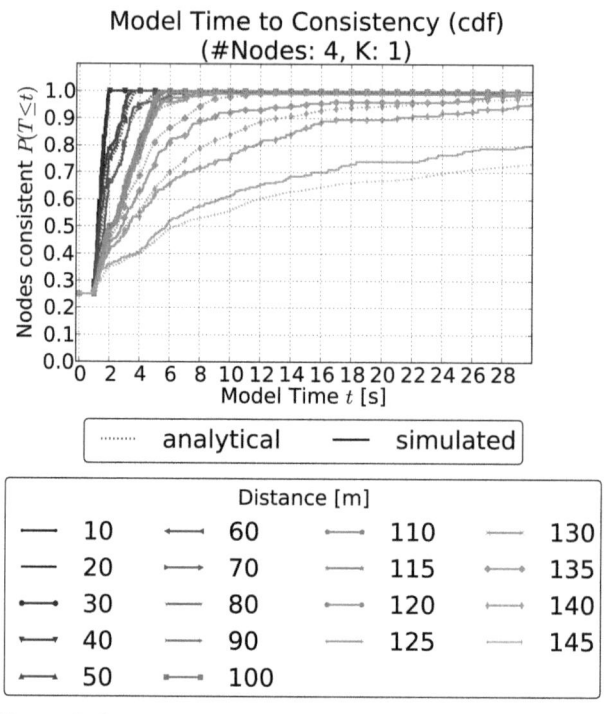

**Figure 9.2:** Network Consistency Delay Distribution (Line-CPM Scenario, 4 Nodes, K=1)

Again, the close fit of the analytical model and the simulated results can be seen for all distances in a Line-CPM scenario for a value of $K = 1$. The fit is especially good for the distances in which WSNs should be operated.

The results from simulation and the analytical model for a 'Line-CPM' scenario with 9 nodes and for $K = 3$, are shown in figure 9.3. It can be seen that the initial jump is reduced to 1/9 because only 1 out of the 9 nodes is injecting the service. Since the overall scenario size is increased and thus the number of hops is increased, the distribution generally consists of distributions of higher convolutions and in general it takes longer until the same consistency level has been reached as for the 4 node scenario in figure 9.1 as expected. Just for the inter-node distance 10 m all nodes are within reach of the initial node and thus lead to the linear increase, while for 4 nodes distances up to 50 m exhibited the linear increase. For distances below 80 m each node has more than 1 successive neighbour. From 80 m

## 9.1 Comparison of Analytical Model and Simulation Results

on there is only 1 successive neighbour (possibly with a PRR < 1) and thus the gap between the distributions for 70 m and 80 m can be explained.

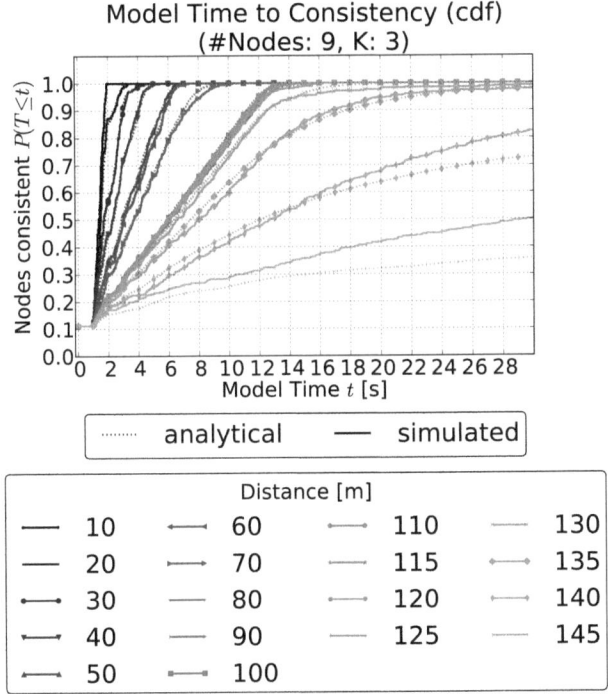

**Figure 9.3:** Network Consistency Delay Distribution
(Line-CPM Scenario, 9 Nodes, K=3)

It can be seen that the analytical model and the simulation results are matching for the increased number of nodes very well except for high distances. At 140 m from 16 s and at 145 m from ca. 12 s onwards a considerable difference between the analytical and simulation results can be seen. This is due to the fact that the combinations imposed by the $K$ value of 3 and the bad link bhaviour at those distances cannot be modeled properly. Note that WSNs should not be operated in this mode as a reliable functioning cannot be guarantueed.

Figures 9.4 and 9.5 show the result for the same 9 node scenario for K values of 1 and 9. The analytical models and the simulation results are closely resembling each other for these cases as well. This validates the model and simulation results with respect to the Trickle parameter $K$. The delay tends to be higher for $K = 1$ compared to $K = 3$ for distance between 50 m and 70 m, where several nodes are in the same transmission/collision domain and thus the parameter $K$ can have an effect. For distances of 80 m and higher no effect can be seen as expected, because each node has exactly one predecessor node.

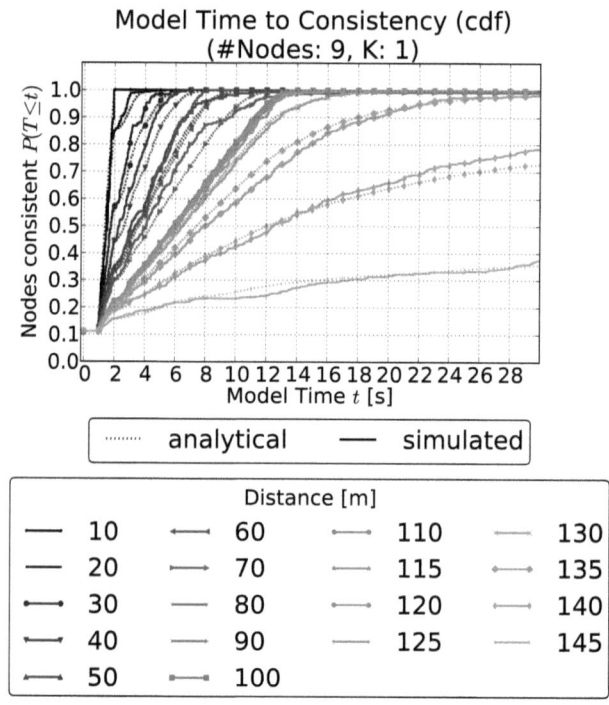

**Figure 9.4:** Network Consistency Delay Distribution (Line-CPM Scenario, 9 Nodes, K=1)

## 9.1 Comparison of Analytical Model and Simulation Results

A higher value of $K = 9$ does not lower the delay compared to $K = 3$ as the scenario size limits the number of nodes in a collision domain for high distances. If distances are low, most nodes are in reach of the initial node, which does not have to compete with other nodes. The remaining informed nodes which have to compete for transmission would in all possible Trickle transmission possibilities inform the remaining uninformed nodes, irrespective of whether $K$ is set to 3 or higher values.

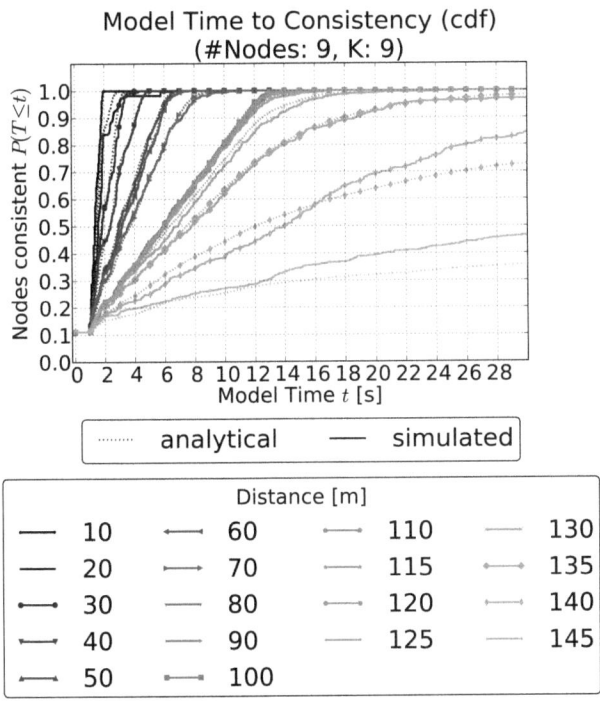

**Figure 9.5:** Network Consistency Delay Distribution (Line-CPM Scenario, 9 Nodes, K=9)

In figures 9.6 and 9.7 the results from the simulations for the three different $K$ values are shown for 9 and 225 nodes in a Line-CPM scenario. It can be seen in figure 9.6 that for a low distance of 10 m there is no difference between the $K$ values (since all nodes are within 1 hop radio range), for higher distances the higher $K$ values of 3 and 9 tend to have faster times until consistency is reached. The difference between $K = 3$ and $K = 9$ is rather small though (while sending considerably more number of messages with $K = 9$). A similar behaviour can be observed for 225 nodes in figure 9.7. For $K = 3$ and $K = 9$ and an inter-node distance of 10 m complete consistency is reached at approximately 22 s, while for the low $K = 1$ complete consistency is reached at about 30 s.

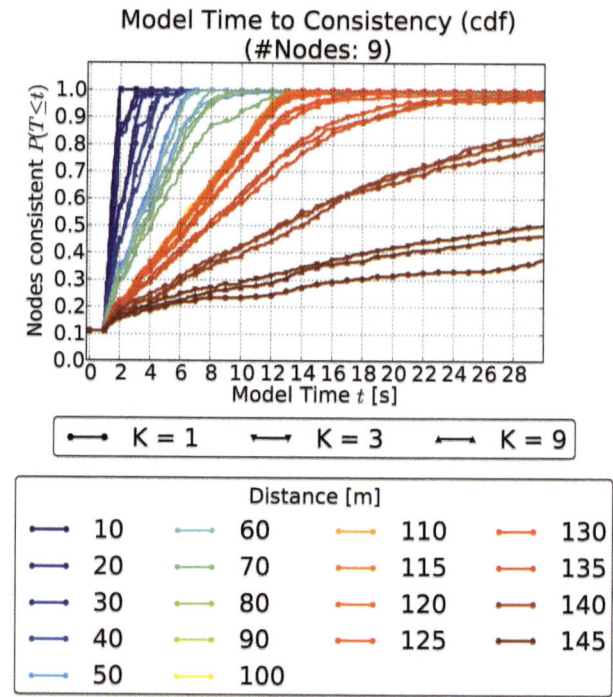

**Figure 9.6:** Network Consistency Delay Distribution (Line-CPM Scenario, 9 Nodes)

## 9.1 Comparison of Analytical Model and Simulation Results

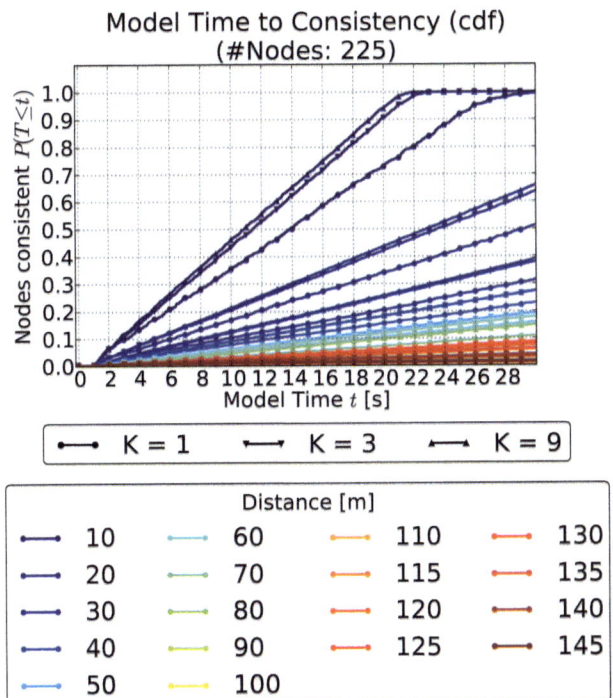

**Figure 9.7:** Network Consistency Delay Distribution (Line-CPM Scenario, 225 Nodes)

In figures 9.8 and 9.9 the delay distribution for the Line-CPM scenario with the number of nodes increased to 16 and 25 and $K = 3$ is shown. As expected, the delay increases with the number of nodes in the scenario. And as for the previous scenarios, the simulation results and the results of the analytical model are fitting closely, thus validating each other with respect to the number of nodes.

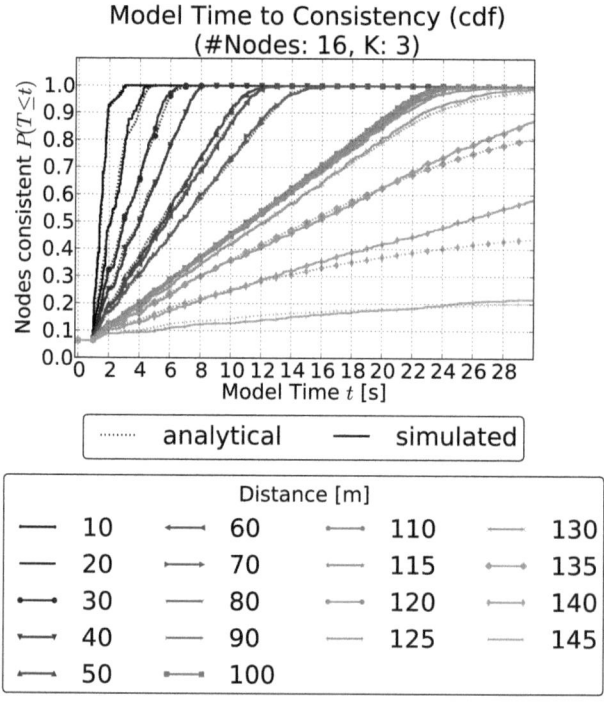

**Figure 9.8: Network Consistency Delay Distribution (Line-CPM Scenario, 16 Nodes, K=3)**

## 9.1 Comparison of Analytical Model and Simulation Results

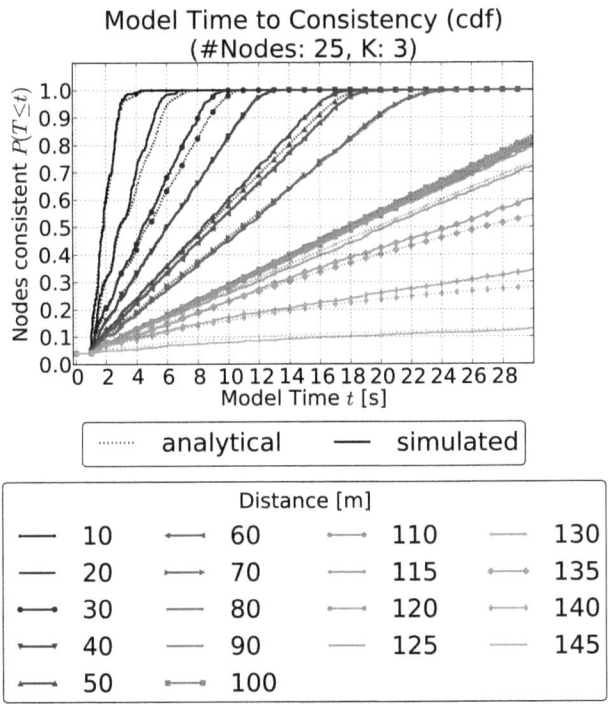

Figure 9.9: **Network Consistency Delay Distribution (Line-CPM Scenario, 25 Nodes, K=3)**

The analytical results and the simulation results for Grid-CPM scenarios with a number of nodes of 4 and 9 are depicted in figures 9.10 and 9.11. These results validate the analytical model for a different scenario type as well.

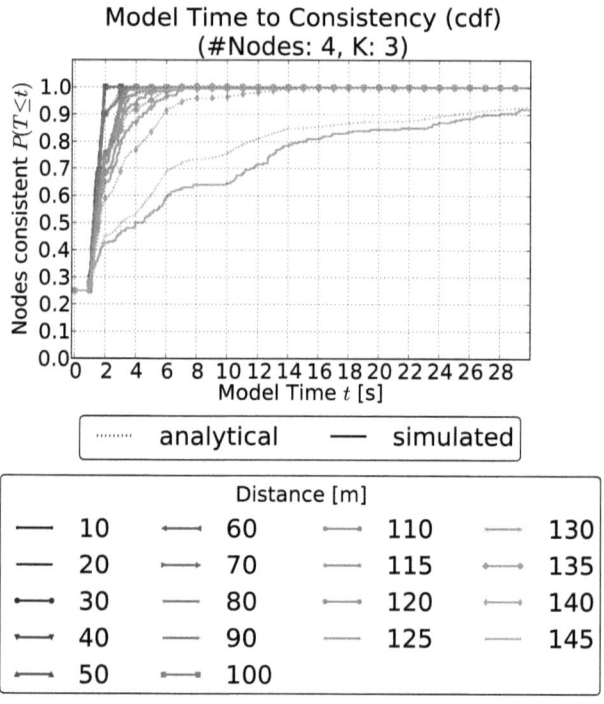

Figure 9.10: Network Consistency Delay Distribution (Grid-CPM Scenario, 4 Nodes, K=3)

By the addition of several convoluted distributions an almost linear increase of the consistency level can be seen for distances from 40 m onwards for scenarios with a small number of nodes or from 10 m onwards for larger scenarios such as the 25 node Line-CPM scenario. A simplified analytical model derived for those distances is detailed in appendix G.

## 9.1 Comparison of Analytical Model and Simulation Results

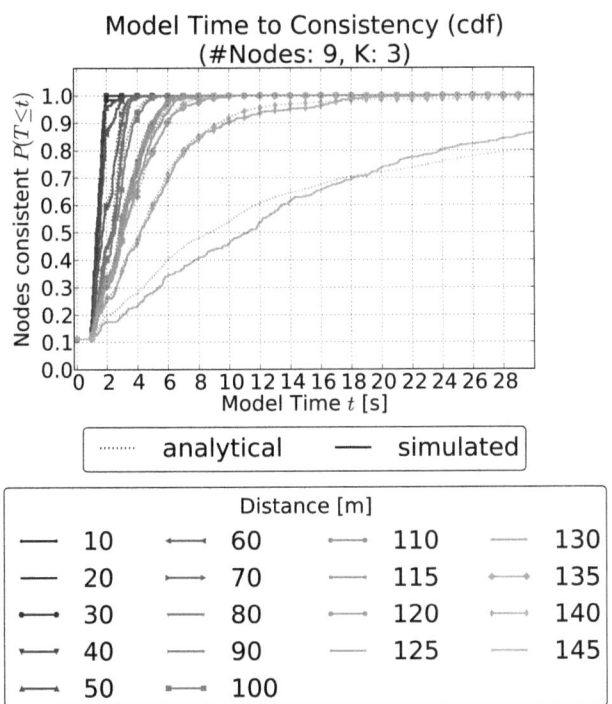

**Figure 9.11:** Network Consistency Delay Distribution (Grid-CPM Scenario, 9 Nodes, K=3)

In figure 9.12 the results from the simulations for the three different $K$ values are shown for 225 nodes in a Grid-CPM scenario. The delay performance is almost identical for $K = 3$ and $K = 9$, while latter is sending considerably more messages. $K = 1$ shows a slower performance for all distances except the most dense scenario.

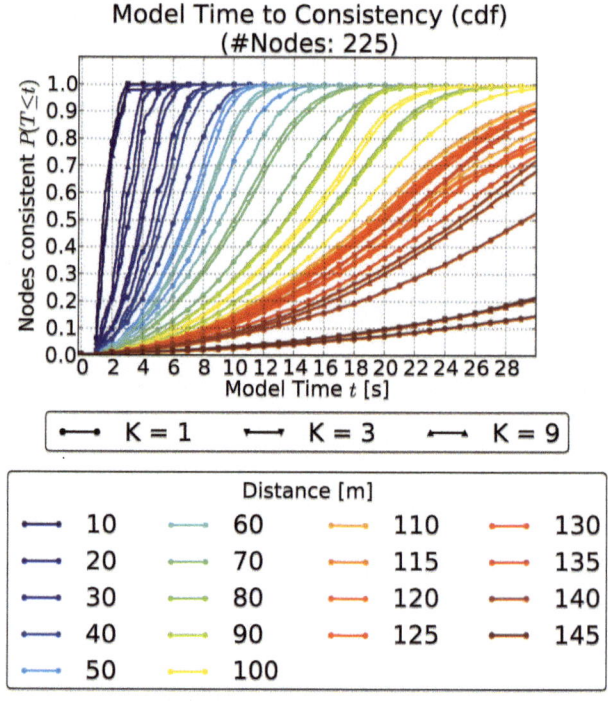

**Figure 9.12:** Network Consistency Delay Distribution (Grid-CPM Scenario, 225 Nodes)

## 9.1 Comparison of Analytical Model and Simulation Results

In figure 9.13 a comparison of the simulation results between the Line-CPM and the Grid-CPM scenario is shown for 9 nodes and various $K$ values. As expected the distribution is faster in the Grid-CPM scenario, as the network diameter is smaller and the nodes have more neighbours.

The same comparison of Line-CPM and Grid-CPM is shown in figure 9.14 for 225 nodes. The faster distribution in a Grid-CPM scenario becomes even more pronounced with higher number of nodes. In a high density Grid-CPM network ($d = 10m$) a complete consistency is reached after approx. 3 s, while it takes approx. 30 s in a Line-CPM network for $K = 1$. A higher $K$ value improves the distribution in a Grid-CPM network only slightly, while in a Line-CPM network a value higher than 1 can yield a considerable improvement. This is due to the fact that a value of 1 can interrupt the rippling of the service through the line, if just one node right behind the consistency border is sending and thus keeping the node at the consistency border from informing new nodes. The increase of $K$ from 3 to 9 yields very little further improvement.

Concludingly, it can be said that the results of the analytical model conform to the expectation, expressed in figure 7.3, and as well to the simulation results of the tool described in section 6.1 with little error for a wide range of inter-node distances, for different scenario types, the parameter $K$ and the number of nodes in the scenario.

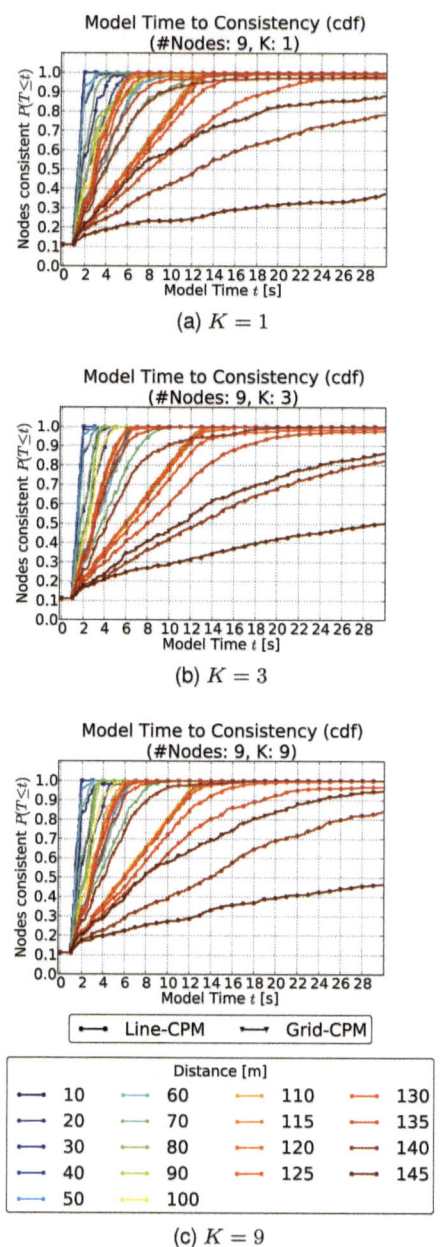

Figure 9.13: Network Consistency Delay Distribution (Simulated, 9 Nodes)

## 9.1 Comparison of Analytical Model and Simulation Results

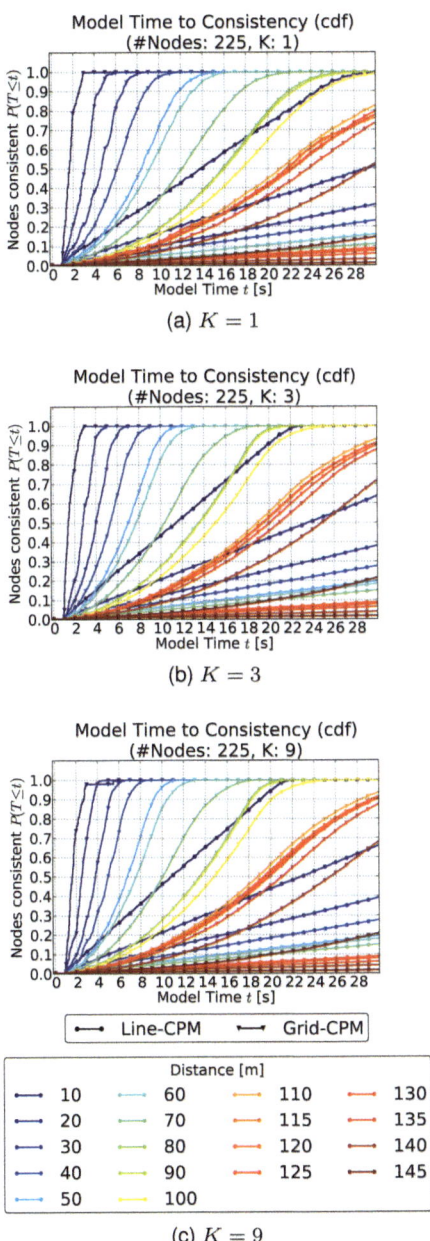

**Figure 9.14: Network Consistency Delay Distribution (Simulated, 225 Nodes)**

## 9.1.2 Mean Number of Sent Packets

In the following sections the results from the analytical model and the simulation for the mean number of sent packets is presented and discussed. The mean number of packets is calculated and measured for 400 s after the injection of a service. The observation time of $400s$ was allowing for the Trickle algorithm to adapt to its lowest rate and distribute to most nodes in all scenarios.

### 9.1.2.1 Mean Number of Packets: Varying K, N=4, Line

Starting with a Line scenario of 4 nodes and varying the value of the Trickle consistency K, the results are shown in figures 9.15, 9.16, and 9.17. The effect of a low, medium and high value of K can clearly be seen.

A low value of K (here 1) allows the algorithm to adapt its transmission rate to the node density of the scenario. For low inter-node distances ($\leq 40$ m, high density) the mean number of sent packets in the observation period is limited to less than 20 sent packets. For high densities a higher mean number of packets is created by the Trickle algorithm. The analytical and simulation results follow the same trend and show similar results. Note, that for scenario distances below 40 m, all nodes in this scenario are in range of each other. While for distances between 80 and 120 m each node can only reach its direct neighbour.

The mean number of sent packets given by the analytical models are slightly lower than the results from the simulation for $K = 1$ due to an underestimation of the active neighborhoods, which results in a too low sending probability for each node. For the higher $K$ values the active is almost the entire neighborhood and thus not underestimated.

With the medium setting of $K = 3$ the ability to adapt to the density of the scenario is restricted, thus more packets are sent, as shown in figure 9.16. For low distances, all nodes are neighbours of each other resulting in 4 nodes in the neighbourhood of each node including itself. Since $K$ is set to 3, 3 out of the 4 nodes can transmit. For distances between 80 m and 140 m each node has only up to 2 neighbouring nodes (directly to the left and right). The fourth node of the line is not a neighbour node of the first node anymore. This results in a neighbourhood size of 3 and therefore Trickle does not restrict the sending with a $K$ value of 3. For the distance 145 m a slight dip is present, which is due to the fact that the propagation model gives a rather high loss rate for the links, so that not all nodes are informed immediately. The analytical model doesn't take this into account.

The simulation typically shows a slightly lower value for the $K$ values which are not 1, since the analytical model assumes that all nodes are starting their Trickle

## 9.1 Comparison of Analytical Model and Simulation Results

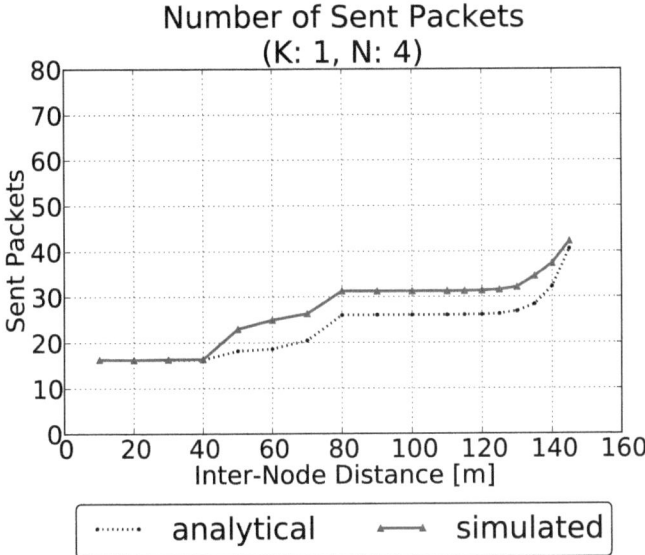

**Figure 9.15:** Mean Number of Sent Packets in a 4 Node Line Scenario (K=1) (Observation Time: 400 s)

cycle immediately at the beginning of the simulation ('late-or-never-informed effect'). Thus, the analytical model is a higher bound here.

A high value of $K = 9$ restricts the ability to adapt to the density even further, so that no adaptation is possible in a 4 node scenario, cf. 9.17.

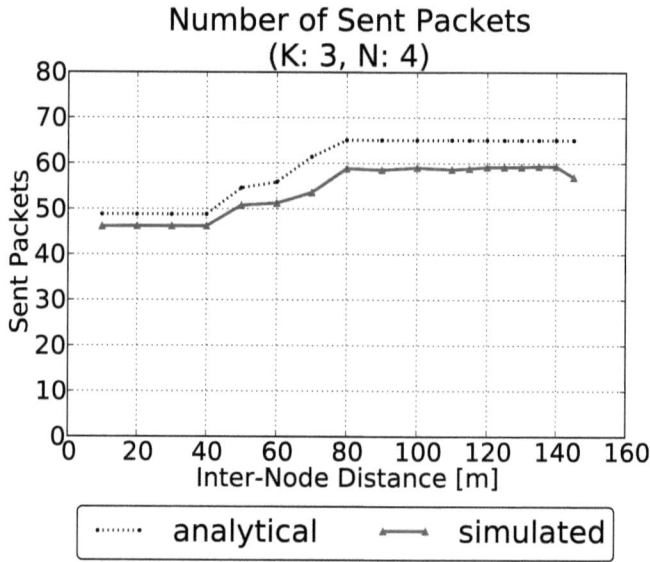

**Figure 9.16:** Mean Number of Sent Packets in a 4 Node Line Scenario (K=3) (Observation Time: 400 s)

## 9.1 Comparison of Analytical Model and Simulation Results

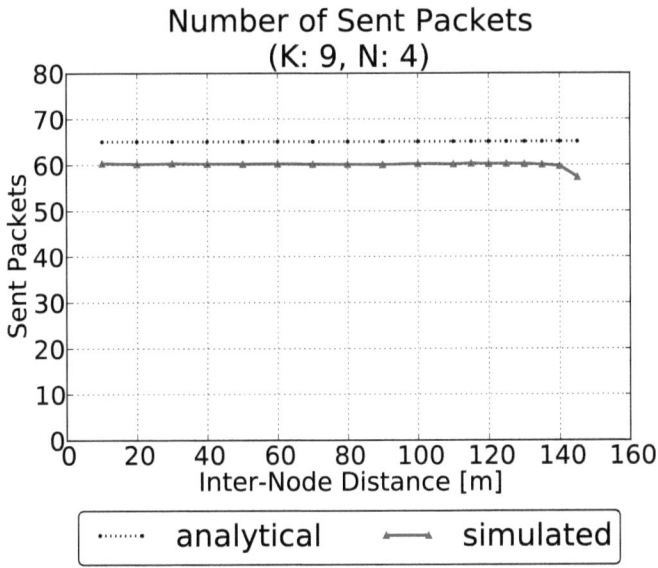

Figure 9.17: **Mean Number of Sent Packets in a 4 Node Line Scenario (K=9) (Observation Time: 400 s)**

### 9.1.2.2 Mean Number of Packets: Varying K, N=64, Line

With an increased number of nodes to 64, the total mean number of packets of all nodes is increased to $\approx 1050$ maximum. The results for the simulation and the analytical model are shown for 64 nodes in figure 9.18 for $K = 1$, in figure 9.19 for $K = 3$, and for $K = 9$ in figure 9.20.

The analytical model shows a similar trend and similar results for all 3 values of $K$ as for the 4 node scenario: With an increased inter-node distance, the number of packets is increased. Again, with a higher $K$ the ability to adapt to the density is restricted. The reasoning for a generally lower number of sent packets for the analytical model compared to the simulated number for $K = 1$ is the same as in the case of 4 nodes. For higher $K$ values the analytical model overestimates the number of packets for most inter-node distances due to the late-or-never-informed effect. For the distances of 140 m and 145 m the simulated number of packets dips because of the nodes are getting even later informed due to the low PRR (0.7 and 0.4) and while the analytical model ignores that fact.

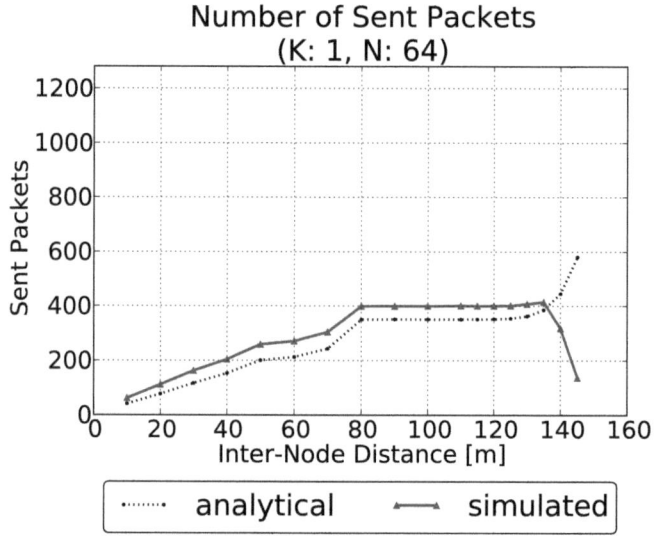

**Figure 9.18: Mean Number of Sent Packets in a 64 Node Line Scenario (K=1) (Observation Time: 400 s)**

9.1 Comparison of Analytical Model and Simulation Results 123

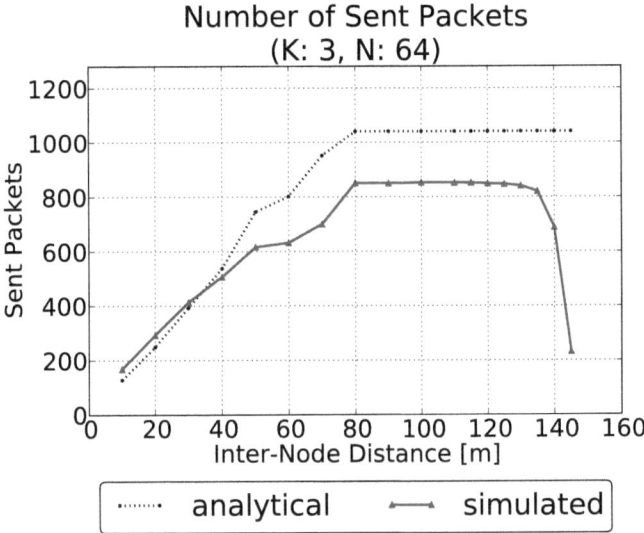

**Figure 9.19:** Mean Number of Sent Packets in a 64 Node Line Scenario (K=3) (Observation Time: 400 s)

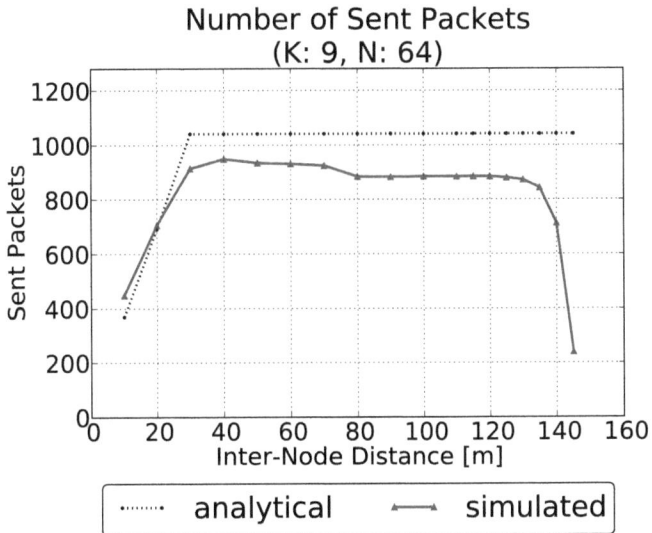

**Figure 9.20:** Mean Number of Sent Packets in a 64 Node Line Scenario (K=9) (Observation Time: 400 s)

### 9.1.2.3 Mean Number of Packets: Varying N, K=1, Line

In figures 9.21, 9.22, and 9.23 the number of nodes in the scenario is varied for a setting of $K = 1$. The mean number of sent packets is mainly effected by the number of nodes.

Again, for high inter-node distances the effect of not or late informed nodes is prevalent for simulation, while that effect is not modeled in the analytical model.

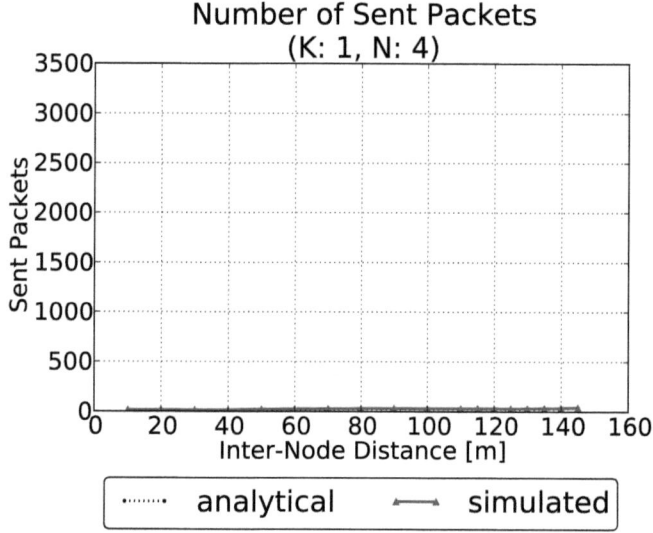

Figure 9.21: Mean Number of Sent Packets in a 4 Node Line Scenario (K=1) (Observation Time: 400 s)

## 9.1 Comparison of Analytical Model and Simulation Results

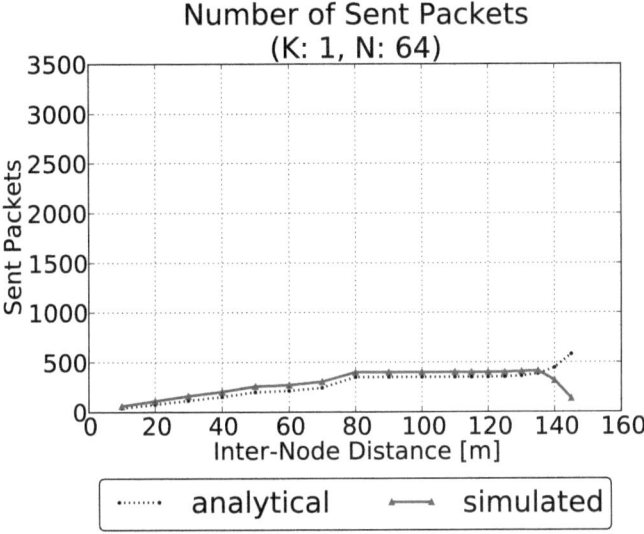

**Figure 9.22:** Mean Number of Sent Packets in a 64 Node Line Scenario (K=1) (Observation Time: 400 s)

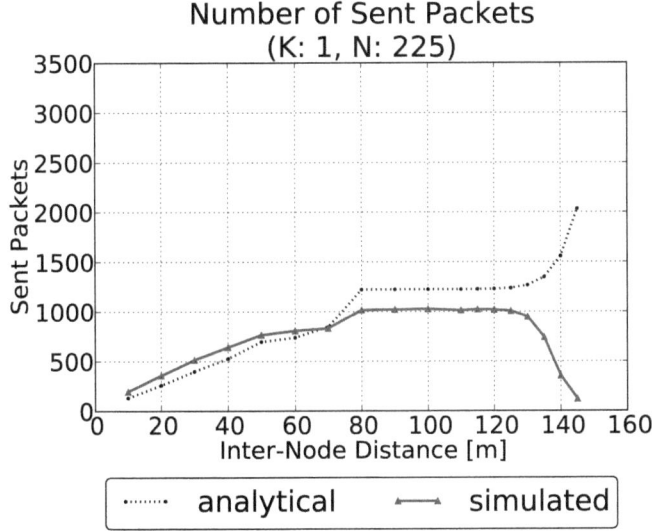

**Figure 9.23:** Mean Number of Sent Packets in a 225 Node Line Scenario (K=1) (Observation Time: 400 s)

### 9.1.2.4 Mean Number of Packets: Varying N, K=3, Line

Varying the number of nodes N with a $K$ value of 3 gives the analytical and simulation results as shown in the figures 9.24, 9.25 and 9.26. Increasing the $K$ value from $K = 1$ to 3, increases the number of sent packets approximately by a factor of 2. The general trend of the curves is staying similar when comparing $K = 1$ and $K = 3$ as well when comparing the different number of nodes. In figure 9.26 the difference between the analytical and the simulation results can again be explained by the late-or-never-informed effect.

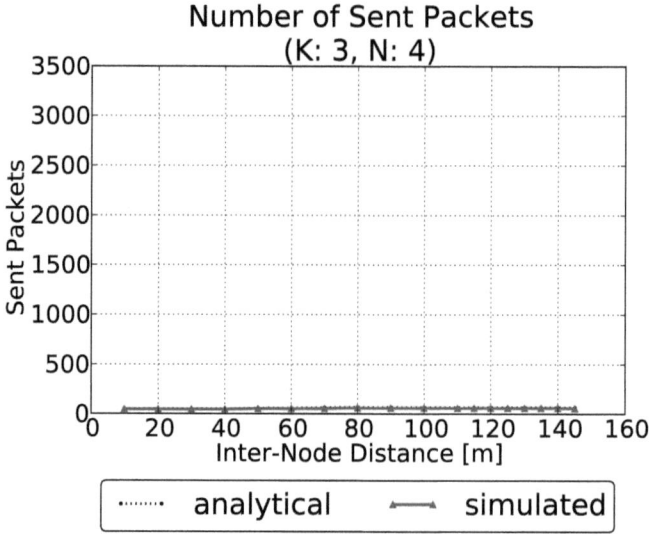

**Figure 9.24: Mean Number of Sent Packets in a 4 Node Line Scenario (K=3) (Observation Time: 400 s)**

## 9.1 Comparison of Analytical Model and Simulation Results

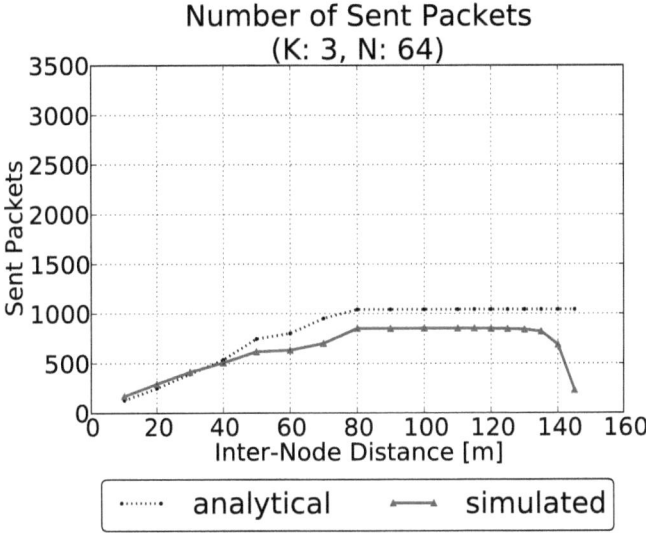

**Figure 9.25:** Mean Number of Sent Packets in a 64 Node Line Scenario (K=3) (Observation Time: 400 s)

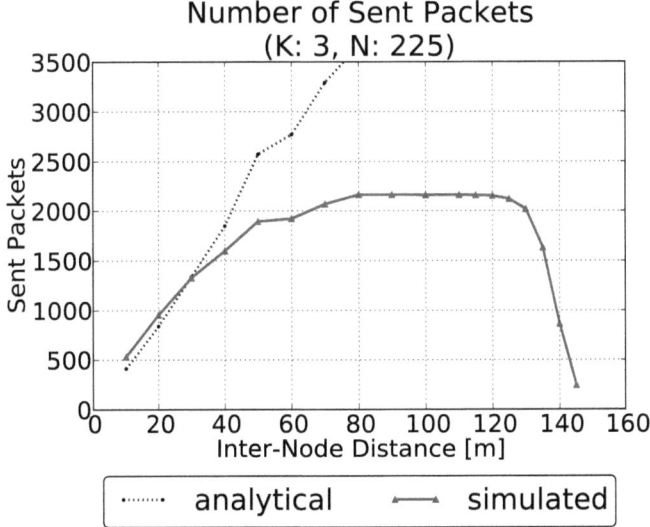

**Figure 9.26:** Mean Number of Sent Packets in a 225 Node Line Scenario (K=9) (Observation Time: 400 s)

### 9.1.2.5 Mean Number of Packets: Line/Grid, N=4, K=3

The graphs for comparing the difference between a Line and a Grid scenario are shown in figure 9.27 and figure 9.28 for 4 nodes and $K = 3$.

As nodes in a grid scenario tend to have more neighbours because of the reduced scenario diameter, the probability of each node sending tends to be lower. The effect can be seen for the 4 node scenario in the distance range between 40 and 110 m. The number of neighbours is reduced between 40 and 80 m for the line scenario, while for the grid scenario the reduction in the number of neighbors starts at 100 m. Recall, that for the line scenario at 80 m only the directly neighbouring nodes can be reached according to the CPM propagation model. At 100 m inter-node distance in the grid scenario the diagonal nodes, which are then 100 m * $\sqrt{2} \approx 142$ m apart, start having non-perfect links.

At low as well as high distances, the number of neighbours are the same in a 4 node scenario for the line respectively the grid scenario, thus the number of sent packets have the same values as well.

9.1 Comparison of Analytical Model and Simulation Results

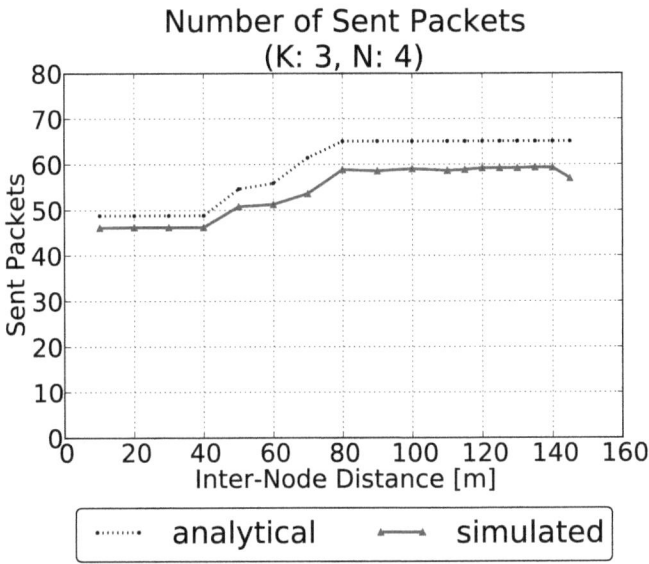

**Figure 9.27:** Mean Number of Sent Packets in a 4 Node Line Scenario (K=3) (Observation Time: 400 s)

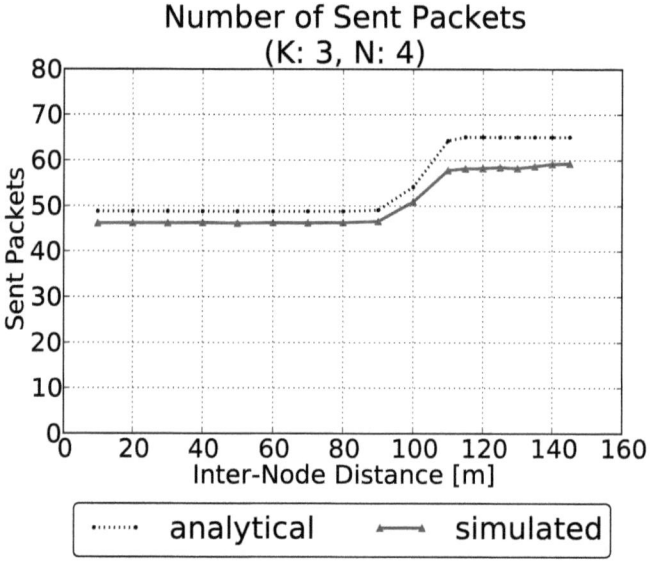

**Figure 9.28:** Mean Number of Sent Packets in a 4 Node Grid Scenario (K=3) (Observation Time: 400 s)

### 9.1.2.6 Mean Number of Packets: Line/Grid, N=64, K=3

The comparison between a line and a grid scenario of 64 nodes is shown in figures 9.29 and 9.30.

As explained earlier, the grid scenario tends to have more neighbours than the line scenario. With the increased number of nodes, this effect is valid for a wider range of inter-node distances. Starting with the inter-node distance of 10 m the line scenario sends more packets than the grid scenario. Only for very high distances, the simulated number of sent packets of the grid scenario is higher than in the line scenario. This is due to the already mentioned effect of late or never informed nodes which are not sending packets. Especially for the line scenario, where each node only has one predecessor node at high distances, the probability of not being informed rises, this can also be confirmed by looking at the delay distribution probabilities in section 9.1.1. The analytical model doesn't take that effect into account, so that the analytical model predicts the same number of sent packets for the line and the grid scenario at an inter-node distance of 145 m.

For most distances the grid scenario exhibits more predecessor neighbours, thus also the probability of each node being informed is increased and reducing the mis-fit of the analytical model for the grid scenario.

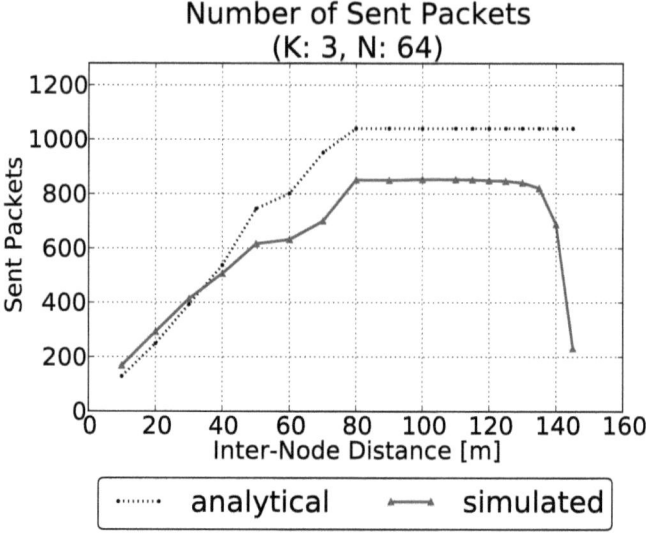

**Figure 9.29:** Mean Number of Sent Packets in a 64 Node Line Scenario (K=3) (Observation Time: 400 s)

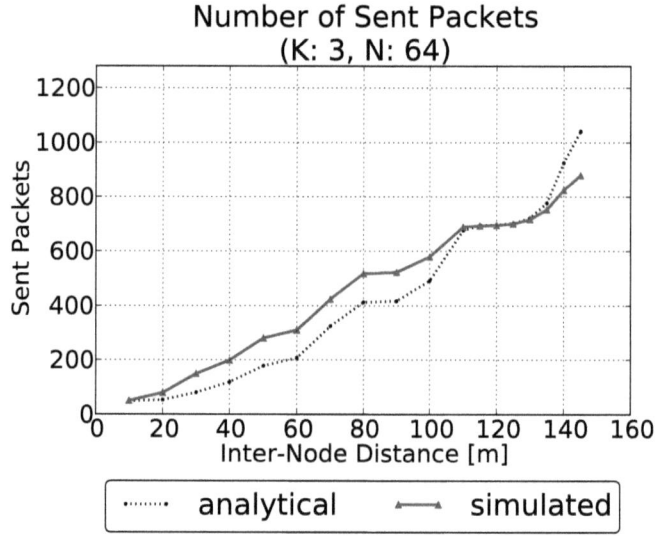

**Figure 9.30:** Mean Number of Sent Packets in a 64 Node Grid Scenario (K=3) (Observation Time: 400 s)

### 9.1.2.7 Mean Number of Packets: Line/Grid, N=225, K=3

For the scenarios with 225 nodes, the results are shown in figure 9.31 for the line scenario and in figure 9.32 for the grid scenario.

The behaviour has not changed much from the 64 node scenarios. The main difference is the increased gap between the analytical model and the simulation results for the line scenario, due to the already mentioned late-or-never-informed effect.

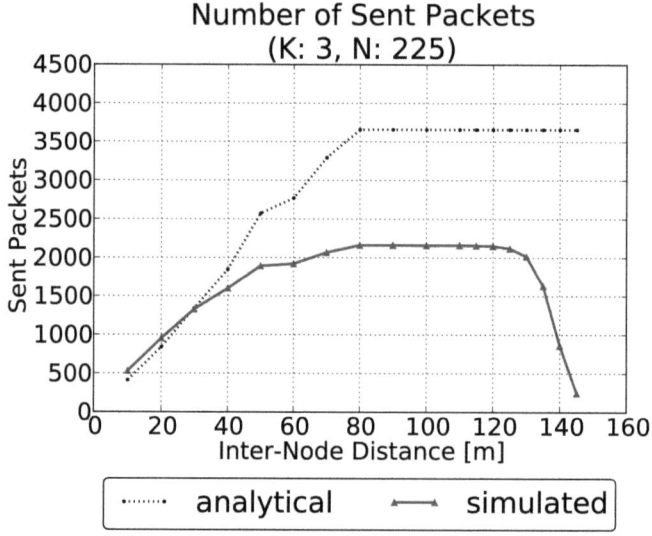

**Figure 9.31:** Mean Number of Sent Packets in a 225 Node Line Scenario (K=3) (Observation Time: 400 s)

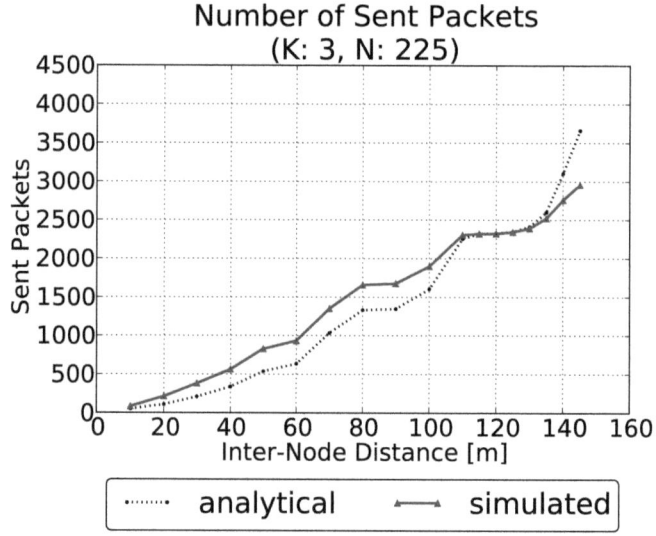

**Figure 9.32:** Mean Number of Sent Packets in a 225 Node Grid Scenario (K=3) (Observation Time: 400 s)

### 9.1.3 Runtime behaviour

One simulation suite run (including evaluation) with 50 Monte Carlo iterations for scenarios with 4, 9, 16, 25, 36, 49, 64, 81, 100, and 225 nodes, and distances of 10, 20, 30, 40, 50, 60, 70, 80, 90, 100, 110, 115, 120, 125, 130, 135, 140, and 145 takes typically approximately 2 days and 6 hours on a simulation server with 16 Intel Xeon CPU X5550 @ 2.67 GHz cores (though only a single core was used) and 48 GB of Random Access Memory (RAM).

For a comparison of the runtime behaviour of the three employed methodological approaches the runtime is tabulated in table 9.2. It can be seen that a simulation with 50 Monte-Carlo iterations is faster than a measurement run with 20 Monte-Carlo iterations. 2.5 times more Monte-Carlo iterations and a 3.5 higher runtime duration in a 8.75 times longer execution of the measurement approach. The analyical approach is 60 times faster than the simulation approach, so approximately equally fast as one single Monte-Carlo simulation iteration. It is to be noted, that the analytical result includes the calculation of base distributions and their convolutions, which can be easily cached. With caching enabled, the analytical model took 33 s to finish and is thus approximately twice as fast as one single simulation iteration.

| Scenario | Simulation | Analytical | Measurement |
|---|---|---|---|
| 9 node Grid | ≈ 1 h (with 50 iterations) | ≈ 50 s | ≈ 3.5 h (with 20 iterations) |

**Table 9.2: Runtime Comparison of the Applied Investigation Methods**

## 9.2 Comparison of Measurement Results with Results from Simulations and the Analytical Model

In order to validate the simulation results as well as the analytical model of the Trickle algorithm, the same implementation that was used for simulation has been executed in the ComNets WSN testbed introduced in section 8.1. In order to closely resemble the link properties measured in section 8.1.1, the simulation propagation model was setup with the link gains as measured in the testbed.

### 9.2.1 Delay

The results of the measurements, the simulations and the analytical model are shown in figures 9.33 to 9.36 for various transmission output power settings. The Trickle parameter settings that have been used are: $\tau_L = 2$s, $\tau_H = 32$s, and $K = 3$.

Starting with a high power setting of 15 (corresponding to -7.00 dBm), the simulated, measured and analytically modelled delay cdfs are depicted in figure 9.33. It can be seen that the simulated (shown with the solid line), the measured results (shown with the dashed line) and the analytical results (shown with the dotted line) are nearly overlapping. The service is distributed to all nodes within 2 seconds, since all nodes are within coverage of the injecting node. The initial value of $1/9 \approx 0.11$ at the start is given by the fact that the injecting node immediately knows about the service.

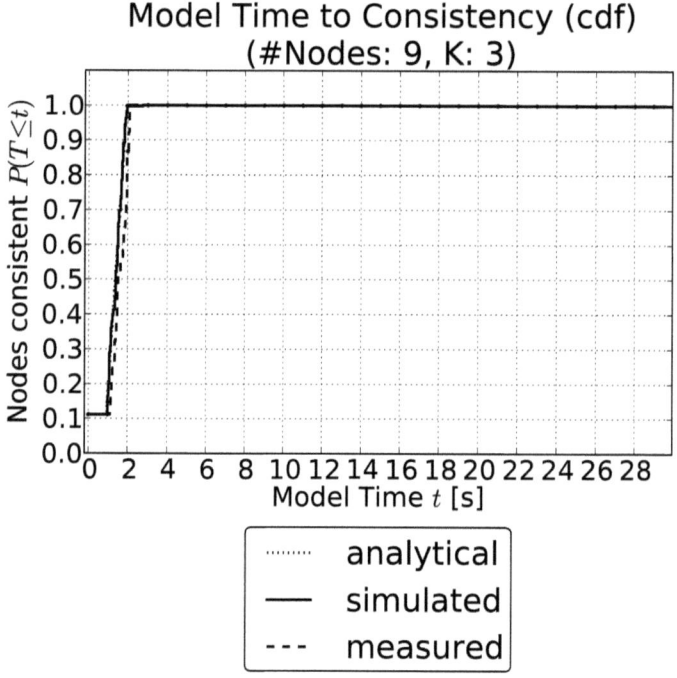

**Figure 9.33: Delay, Power Setting 15**

In figure 9.34 the delay cdf is shown for a transmission output power setting of 3 (or -25 dBm). Again, simulated, modelled and measured cdfs are mostly congruent

## 9.2 Comparison of Measurement Results with Simulations and Analytical Model

and show a very similar behaviour. The delay to reach all nodes is below 4 seconds. The topology of the power setting 3 has been shown in figure 8.6.

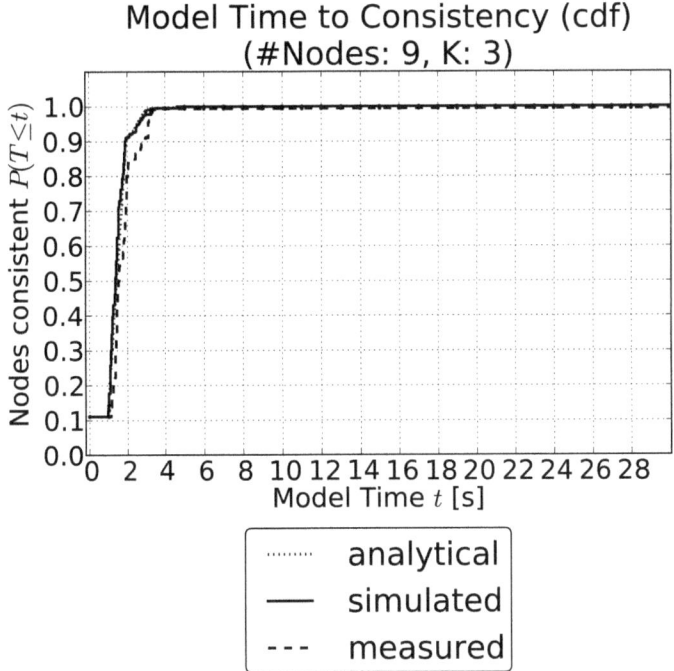

Figure 9.34: Delay, Power Setting 3

When setting the output power level to 2 (-28 dBm), the consistency delay is increased considerably as displayed in figure 9.35. The measured, modelled and simulated results are again reasonably close to each other and show an overall similar shape. The topology for the power setting 2 has been shown in figure 8.5.

The results of a power level of 1 (-33 dBm) are shown in figure 9.36. The delay is further increased with a reasonable match between simulation, analytical model and measured results. In figure 8.4 the topology for the power setting 1 has been shown.

Comparing the power-adapted measured results of figures 9.33 to 9.36 with the non-power adapted simulation results which are shown in figure 9.11 can give an equivalence of the power level and the true inter-node scenario distance that can be matched with the power level. The equivalence is tabulated in table 9.3.

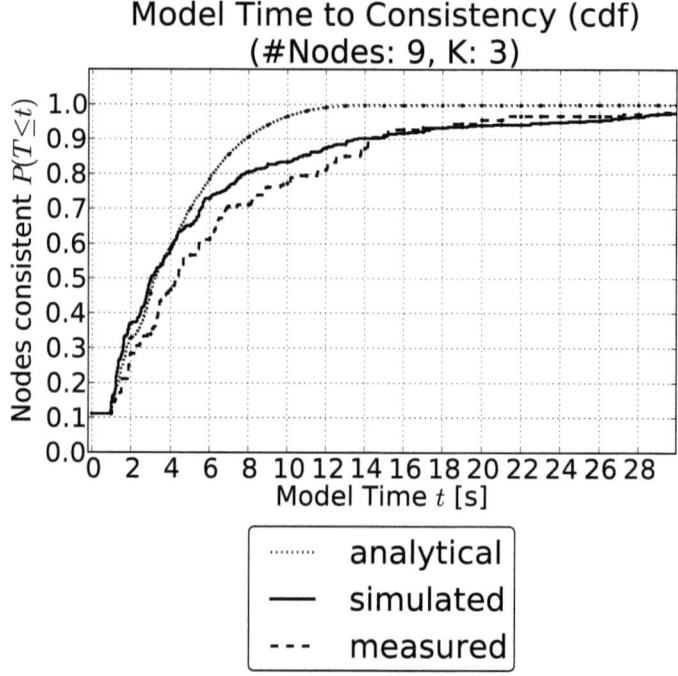

**Figure 9.35:** Delay, Power Setting 2

With the low inter-node distance in the testbed of only 1.2 m and the power level setting restricted to integer values the granularity is that high that not all inter-node scenario distances can be matched. However, the range from very close inter-node distance over several medium distances to high distances can be adequately measured with the small-scale testbed.

9.2 Comparison of Measurement Results with Simulations and Analytical Model 139

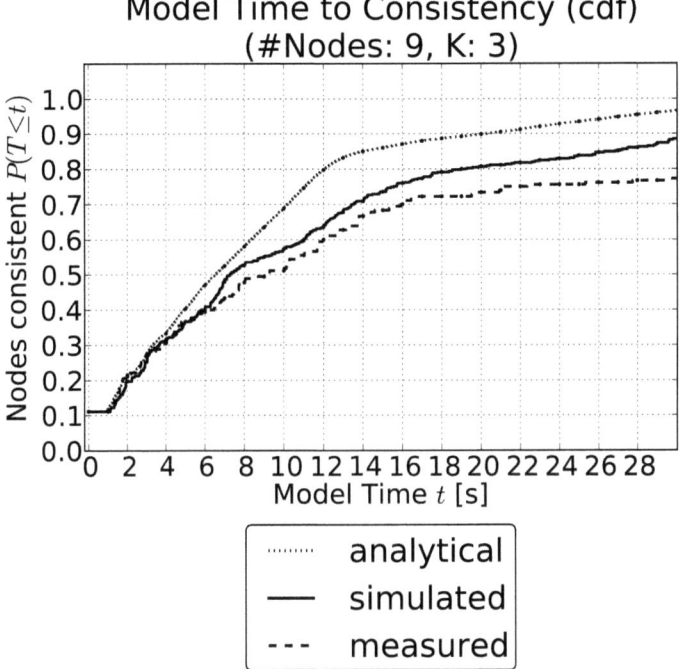

Figure 9.36: Delay, Power Setting 1

| Power Level | Inter-node Distance |
|---|---|
| 15 | 10 m |
| 3 | 50 – 60 m |
| 2 | 135 – 140 m |
| 1 | 145 m |
| 0 | >145 m |

Table 9.3: Equivalence of Testbed Power Level and Inter-Node Distance in Simulation and Analytical Model

## 9.2.2 Mean Number of Sent Packets

The mean number of sent packets that has been observed in the measurement runs as well as in the simulation runs and the analytical model is shown in table 9.4 for various power levels and the same settings as in the previous section.

A good match between the simulated, modelled and measured mean number of packets can be seen. For the lowest power level 0, where no consistency can be reached as shown in the previous section, a low mean number of packets is sent (mostly by the injecting node only). The analytical model considerably overestimates the number of sent packets for the lowest power level, due to neglecting the late-or-never-informed effect. For the measured and simulated resuts, the power level 1 has the highest mean number of sent packets, due to its sparse connectivity. With higher power levels the mean number of sent packets is reduced, since the network is more dense and the Trickle algorithm adapts to the density.

| Power Level | Simulated | Measured | Analytical Model |
|---|---|---|---|
| 0 | 15.3 | 21.0 | 146.4 |
| 1 | 129.4 | 118.9 | 122.4 |
| 2 | 126.7 | 114.5 | 84.1 |
| 3 | 85.6 | 57.4 | 49.0 |
| 15 | 47.6 | 51.9 | 48.8 |

**Table 9.4: Comparison of the Mean Number of Sent Packets**

# 10 Conclusions and Outlook

## 10.1 Conclusions

The presented research work shows the validity of the service concept for Wireless Sensor Networks (WSNs) and introduces an implementation of this concept for multi-hop networks. The feasability of the exchange of services between IP and Sensor Networks has been demonstrated. The implementation has been used in various demonstrations at workshops.

Possible distribution algorithms (Regular Inteval Pushing and Trickle) embedded in the service framework have been surveyed, implemented, simulated, measured, analysed, and compared.

The Trickle algorithm has been extended to support the detection of services that have vanished and do not need to be distributed anymore.

A simulation tool has been extended to enable the simulation of the algorithms. Various scenarios have been created to study the behaviour for possible application areas. By means of simulation the algorithms were evaluated. The statistical significance of the results was established by using a Kaplan-Meier estimator which estimated the confidence interval by using Greenwood's formula.

The mostly uniform spatial distribution and edge effects of the Trickle algorithm have been shown. Nodes with fewer neighbours can have twice the sending rate compared to nodes with an average number of neighbours. The influence of the Trickle parameters has been analysed. While $\tau_H$ influences the steady-state sending rate, $\tau_L$ mostly influences the distribution delay. The parameter $K$ influences the sending rate in the Trickle distribution phase as well as in the steady-state phase.

The optimised selection of the Trickle parameters which are based on application requirements for specific scenarios has been shown by means of tabulated 95 delay percentiles. It has been shown that medium $K$ values show a better delay performance than low or high $K$ values. Obviously, Line scenarios have higher delays than Grid scenarios of the same inter-node distance, since latter ones have more number of neighbours and can distribute in more than one direction.

Analytical models for the delay cdf of the Trickle algorithm and the mean number of packets sent by the Trickle algorithm have been created in this thesis. The analytical model for the delay of the Trickle algorithm is based on combinations of convolved unit pdfs. The combination factor of the pdfs is derived from the scenario description. The model for the number of sent packets is derived from the Trickle cycles and average number of neighbours.

Additionally to the simulations and analytical modelling, measurements with typical WSN hardware have been performed in a dedicated 9 node WSN ceiling testbed at the Communication Networks group of the University of Bremen. The resulting link conditions have been measured and the implementation that has been used for simulations has also been deployed and observed in the testbed. Even with the small spatial dimensions, bigger scenarios could be evaluated by reducing the power level, e.g. with a power level of 3 scenarios of 50-60 m inter-node distance could be matched.

All three scientific methodologies that have been employed in this thesis (simulations, analytical modelling, and measurements) were used to gather results on delay of service distribution and number of sent packets of the Trickle algorithm. The similarity of the results from all three methods has been shown for different scenarios over a wide range of inter-node distances and for various settings of the algorithm and can thus be relied on.

The analytical model derived in this thesis, can deliver the results in about $1/60^{th}$ of the time of the simulation (when run with 50 Monte-Carlo iterations) for a 9 node Grid scenario. The simulation is about 3.5 times faster than the measurements (with only 20 iterations).

The results show that the Trickle algorithm is an adaquate algorithm for service discovery which adapts nicely to various network densities and can be parameterised to match application requirements. It is fulfilling the stated requirements and is superior to other algorithms, such as the Regular Interval Pushing which are not adapting to the network density and can match Trickle's performance either for the delay or for the number of sent packets, but not both.

Since the Trickle algorithm is also used in other applications, such as the Internet Engineering Task Force routing protocol RPL, the results obtained in this thesis are not limited to service distribution, but can be generalised. For example, the results can be applied to parameterise for certain route discovery times when using RPL.

The results shown in the thesis are for the distribution of a single service. When the distribution of multiple services is desired, one instance of Trickle can be run for each service. Then, the number of packets generated is just the number of packets sent for the single service case multiplied with the number of services. The delay characteristics is expected to behave similar for each individual service as for the single service case. This has been demonstrated to be feasible in a prototype implementation.

The service framework which has been created in this thesis solves problems that are prevalent in many WSN application areas; logistics is only one of them. The results that have been deduced for the Trickle algorithm can be exploited generally where the IETF WSN stack with RPL is deployed.

## 10.2 Outlook

The results shown in this thesis for a single service can be extended for multiple services in a network in various other ways than implemented in this thesis with multiple Trickle instances. A different option to support multiple services is to generate hashes over multiple services and distribute the hashes by using one instance of the Trickle algorithm. When a new service is added, a new hash over all services is generated and distributed with the new service description.

The presented service distribution algorithm can also be used in conjuction with the Constrained Application Protocol (CoAP) and the resource discovery it is offering. The service distribution can be used for the discovery of CoAP endpoints, the resources (which are offering services) of the CoAP endpoint can then be discovered by using the CoAP resource discovery.

In this thesis, results for the delay and the number of sent packets have been shown. The number of sent packets mainly influences the amount of energy spent by the WSN nodes. Thus, it is a good indicator for the lifetime of the nodes. A more detailed study which uses power simulation extensions of the simulation tool that has been used in this thesis should be performed to be able to give lifetime estimations of nodes in the network as well as of the complete network.

The delay results of the Trickle algorithm can be approximated in many cases by a step-wise linear function. A simplified model could be derived which offers an almost instant estimation of the delay. A first version of such an approximating step-wise linear model is shown in appendix G, which does not take the influence of the parameter $K$ into account, but already gives a good fit to the results of the regular analytical model and simulations.

# A Other contributions to communication networks research

Several other contributions to research in the field of communication networks have been made by the author during the work on this thesis.
In the course of the research projects

- CRC 637 'Autonomous Cooperating Logistic Processes − A Paradigm Shift and its Limitations' subprojects B3 (Mobile Communication Networks and Models (2004-2007)) funded by the Deutsche Forschungsgemeinschaft (DFG),
- DFG funded CRC 637 'Autonomous Cooperating Logistic Processes − A Paradigm Shift and its Limitations' subproject T4 (Monitoring Technologies for Food Transports (2008-2009)) funded by the DFG, and
- 'The Intelligent Container: Linked Intelligent Objects in Logistics (2010-2013)' funded by the BMBF

several deployments of WSNs have been planned, performed and analysed in cooperation with industrial and academic partners. The author deployed together with the project partners WSNs in commercial transport vehicles, in a warehouse in Germany and in a cargo container, which was shipped on a vessel from Central America to Germany. In the deployments the WSN was attached to the Internet by different communication means: by Universal Mobile Telecommunication System (UMTS) and Wireless Local Area Network (WLAN) in the vehicle, by WLAN provided over a UMTS-WLAN router in the warehouse or by WLAN and satellite links on the vessel. For those deployments debugging tools and tools for analysing the measured data have been implemented and used. Results from the deployments have been published in [Bec+08a; Bec+09; Yua+09; Bec+10b; Jed+10a; Jed+11].

Additionally, further study items in the sub-project B3 that have been published are: Communication in Logistics [BT05; Bec+06e], Wireless VoIP [BG06], Demonstrator for Self-Organising Logistics [Bec+06a; Beh+06b; Beh+06a; Beh+06c], Self-Organisation in ICT [Bec+07a; BTG07], Self-Organisation Scenarios [Wen+07; Jed+07] as well as mobile software agents [Bec+06b; Bec+06c; Bec+07b].

Various research items of the author regarding simulation tools for communication networks are published in [BTG06; Bec+08b] (simulation tool comparisons), [Bec+06d; Geh+06] (agent-based and discrete event simulation), [BWG06] (combined logistic and communication simulation). Some more publications have been created in the context of the European Cooperation in Science and Technology (COST) action 285 'Modelling and Simulation Tools for Research in Emerging Multiservice Telecommunications', cf. [Bec+07c] (improved PRNG for the OPNET simulation tool) and [RB08] (a PRNG WebService).

In cooperation with CRC 637 subproject B1 'Autonomously Controlled Routing in Transport Networks' the applicability of the Border Gateway Protocol (BGP) to routing in logistics was studied in [WBG05].

One of the main operating systems for WSNs named TinyOS has been ported to a new hardware platform which is based on the Software Defined Radio Universal Software Radio Peripheral (USRP). The result of this effort is called TinyOS Software Defined Radio (TOSSDR) and has been published in [Bec+10a]. TOSSDR enables physical and medium access control layer research.

In combination with 6LoWPAN WSNs demonstrations of Mobile IP enabled constrained WSN nodes have been sucessfully set up.

Several publications have surveyed recent IETF protocols in the WSN domain [BKG10; BKG11b; BKG11a].

In the BMBF funded 'Intelligent Container' project, the author developed an implementation of the CoAP for TinyOS based on libcoap [Ber10], which has been accepted for inclusion into the main source code repository [Tin12] and is part of the TinyOS 2.1.2 release. This work completes the implementation of a full IETF protocol stack for TinyOS as shown in figure A.1. The author of this thesis contributed the implementation of an HTTP/CoAP-Proxy to libcoap. Publications related to those efforts are [Bec+11; Kul+11].

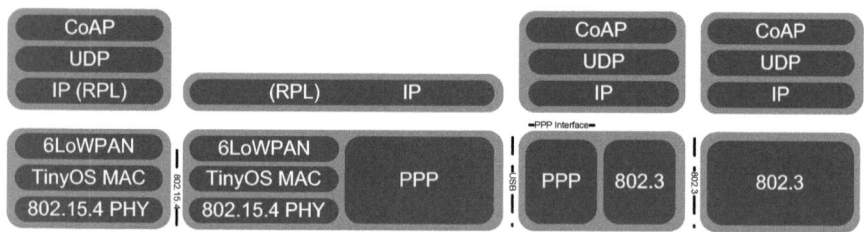

**Figure A.1: IETF Protocol Stack for TinyOS**

On top of that, the IETF draft 'Transport of CoAP over SMS and GPRS' [Bec+13] has been published, which details how to employ CoAP on non-IP or in mixed IP/non-IP networks. Those ideas have been picked up and are currently integrated into European Telecommunications Standards Institute (ETSI) and Open Mobile Alliance (OMA) Lightweight M2M standardisation documents.

Measurements for various popular WSN hardware platforms have been performed and published in [TBG11].

Within the European Union (EU) funded Network of Excellence (NoE) project CRUISE, publications regarding WSN architectures and WSN simulations have been created [Tim+07b; Tim+07a].

A Continuous Integration (CI) setup which permanently monitors the compilability and function of source code for WSNs on real sensor hardware has been shown in [PBG13].

# B  The Minimum of Several Random Variables

Assuming $X_i$ are $n$ independent random variables, then it holds that

$$P(\bigcap_{\forall i}\{X_i > x\}) = \prod_{\forall i} P(X_i > x). \tag{B.1}$$

The minimum of the random variables can then be calculated by

$$\begin{aligned}
P(\min(X_1, X_2, \cdots X_n) \leq x) &= \\
1 - P(\min(X_1, X_2, \cdots X_n) > x) &= \\
1 - P(\bigcap_{\forall i}\{X_i > x\}) &= \\
1 - \prod_{\forall i} P(X_i > x) &= \\
1 - \prod_{\forall i}(1 - P(X_i \leq x)).
\end{aligned} \tag{B.2}$$

If all $n$ random variables are identically distributed, this becomes

$$\begin{aligned}
P(\min(X_1, X_2, \cdots X_n) \leq x) &= \\
1 - (1 - P(X \leq x))^n.
\end{aligned} \tag{B.3}$$

Derivation adapted from [Kre03].

# C The Kaplan-Meier Estimator

The Kaplan-Meier estimator [KM58] estimates survival functions, e.g. from lifetime data of medical illnesses [KP80]. A survival function is a Complementary Cumulative Distribution Function (ccdf), e.g. $S(t)$. The Kaplan-Meier estimator is a maximum-likelihood estimator.

$$\hat{S}(t) = \prod_{t_i \leq t} \frac{n_i - d_i}{n_i} = \prod_{t_i \leq t} \left(1 - \frac{d_i}{n_i}\right) \qquad (C.1)$$

defines the Kaplan-Meier estimator of $S(t)$. The initial condition is $\hat{S}(0) = 1$. $d_i$ is the number of deaths at time $t_i$ and $n_i$ gives the number of survivors just prior to $t_i$. In this thesis a 'death' is equivalent to a node getting informed about a service. A 'survivor' is a node which has not been informed about a service yet. The times of a node getting informed are taken from the Monte-Carlo simulation runs.

Greenwood's formula approximates the variance of the Kaplan-Meier estimator at $t_k \leq t \leq t_{k+1}$:

$$\mathrm{var}\{\hat{S}(t)\} \approx [\hat{S}(t)]^2 \left\{ \sum_{i=1}^{k} \frac{d_i}{n_i(n_i - d_i)} \right\} \qquad (C.2)$$

The resulting $\alpha$-level confidence interval is then given by

$$\left[\hat{S}(t) - w \cdot (\mathrm{var}\{\hat{S}(t)\})^{\frac{1}{2}}; \hat{S}(t) + w \cdot (\mathrm{var}\{\hat{S}(t)\})^{\frac{1}{2}}\right], \qquad (C.3)$$

where $w$ is the multiple of the standard deviation $\sigma$ of the normal distribution that contains values with the probability $\alpha$. The width of the interval is shown for various values of $\alpha$ in table C.1.

| Confidence level $\alpha$ | |
|---|---|
| $\phi(+w\sigma) - \phi(-w\sigma)$ | $w$ |
| 0.80 | 1.281551565545 |
| 0.90 | 1.644853626951 |
| 0.95 | 1.959963984540 |
| 0.98 | 2.326347874041 |
| 0.99 | 2.575829303549 |

**Table C.1: Confidence Level and Approximated Width of the Confidence Interval**

In section 6.3.1 the Kaplan-Meier estimator and Greenwood's formula have been used to derive the confidence interval of the delay cdfs from the ccdfs.

$$S(t) = P[T > t] = 1 - P[T \leq t] \qquad (C.4)$$

The confidence interval of equation C.3 is shown in figures 6.1 and 6.2 for a confidence level $\alpha = 0.95$.

# D  Simulated PRR Topologies for a 9 Node Line-CPM Scenario

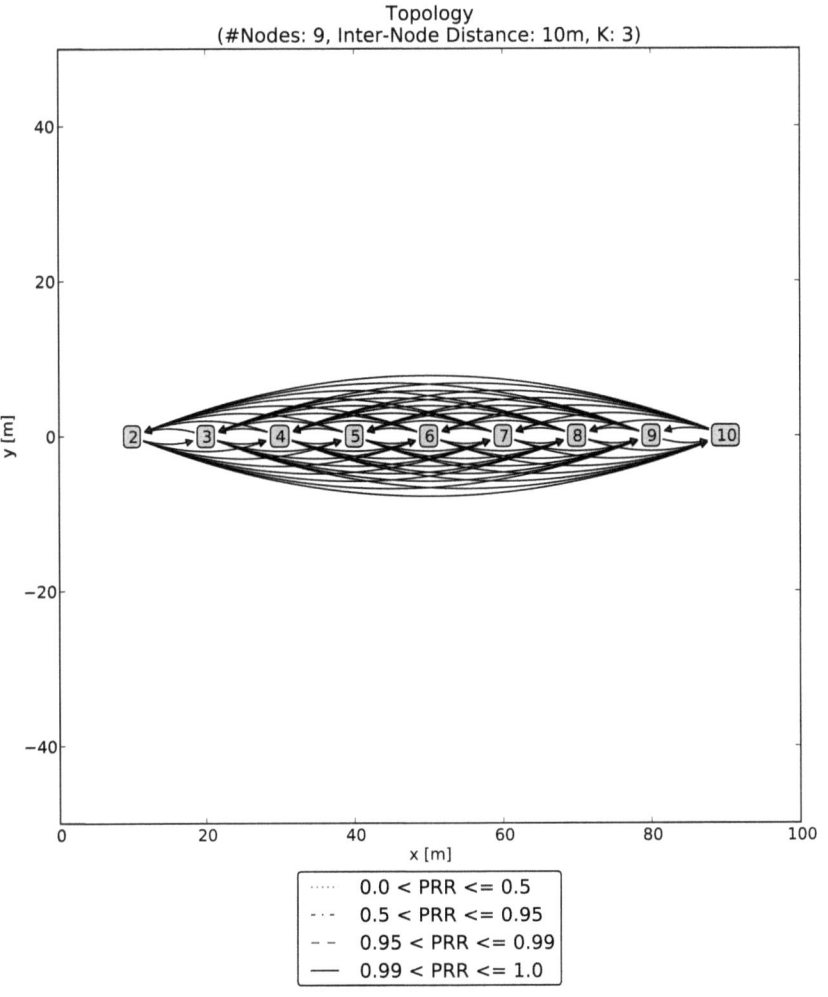

Figure D.1: PRR Matrix (Distance 10m)

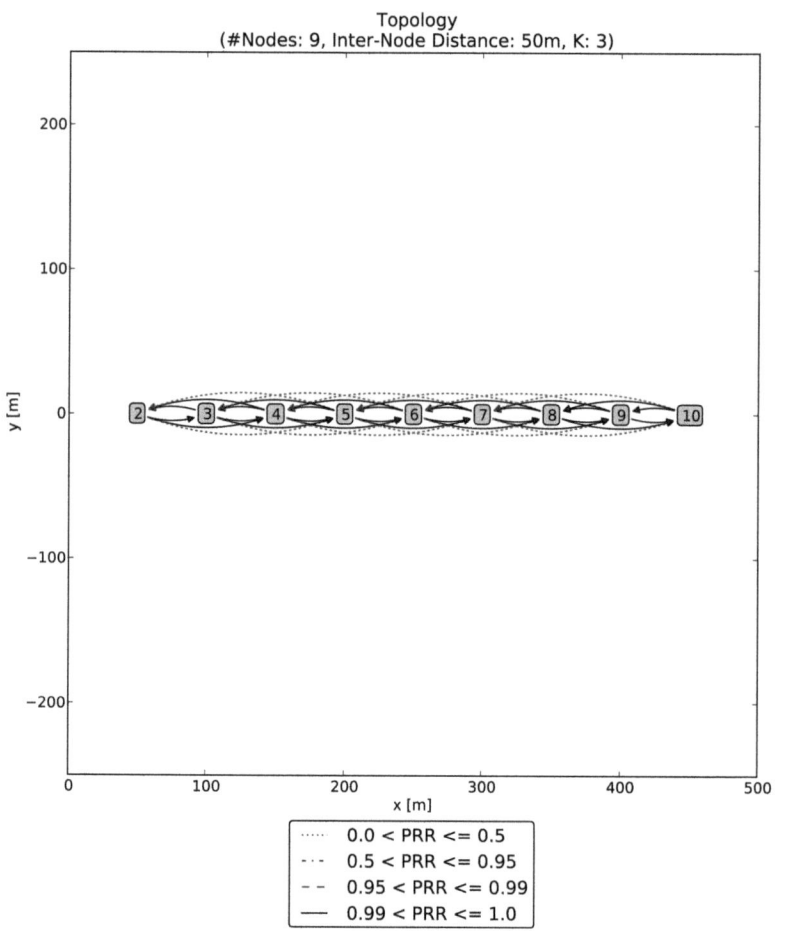

**Figure D.2: PRR Matrix (Distance 50m)**

# Simulated PRR Topologies (Line-CPM)

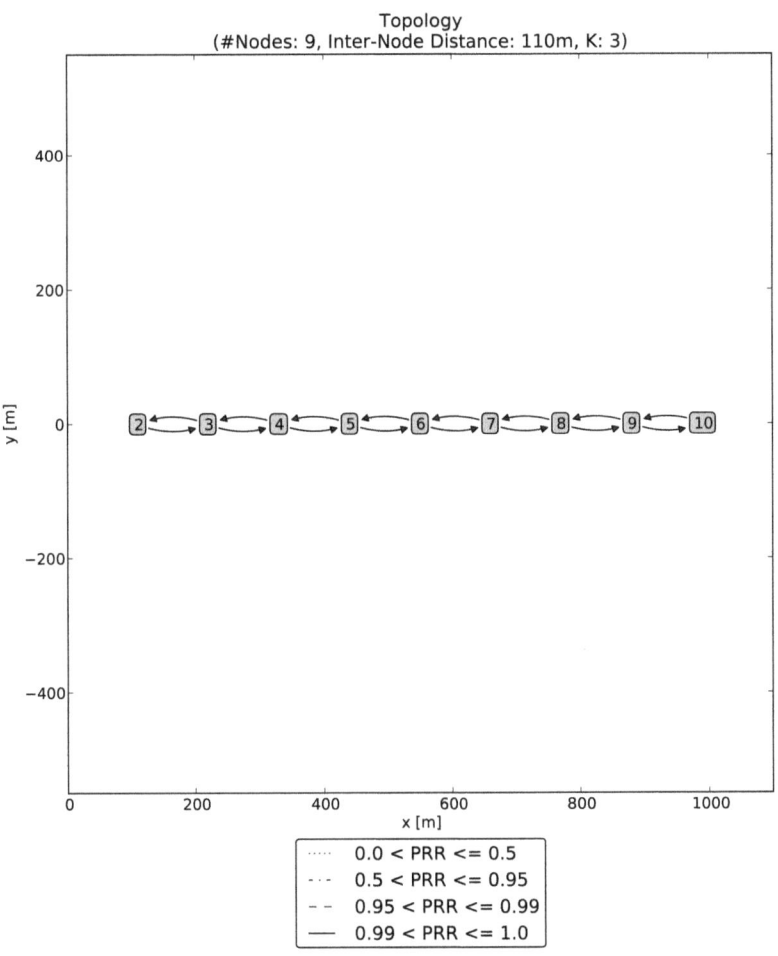

**Figure D.3: PRR Matrix (Distance 110m)**

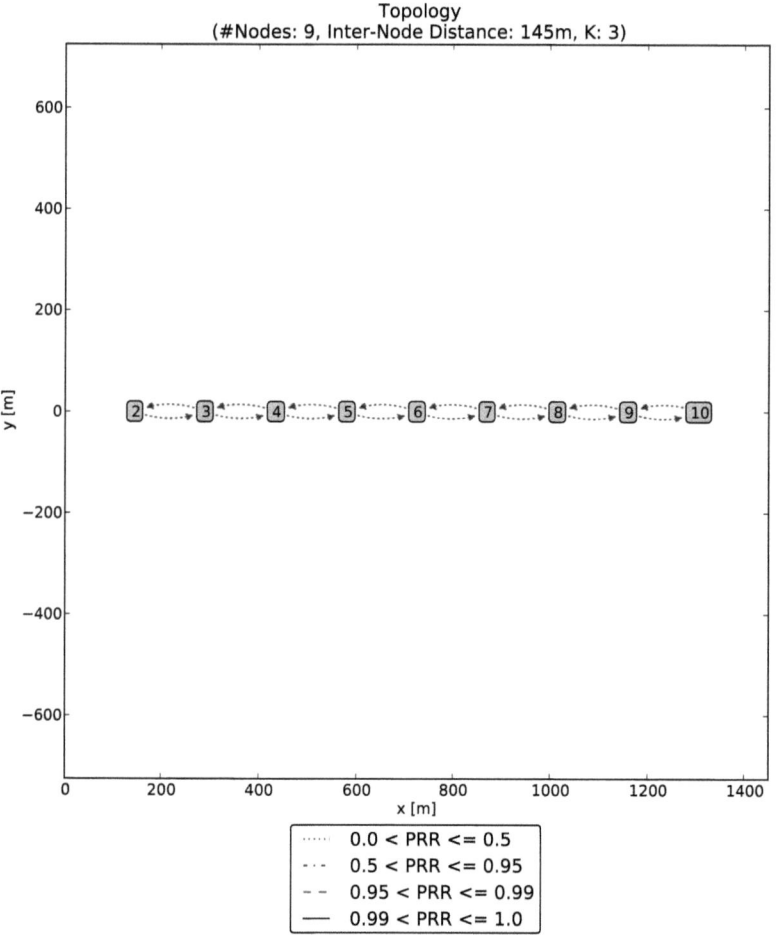

**Figure D.4: PRR Matrix (Distance 145m)**

# E Simulated PRR Topologies for a 9 Node Grid-CPM Scenario

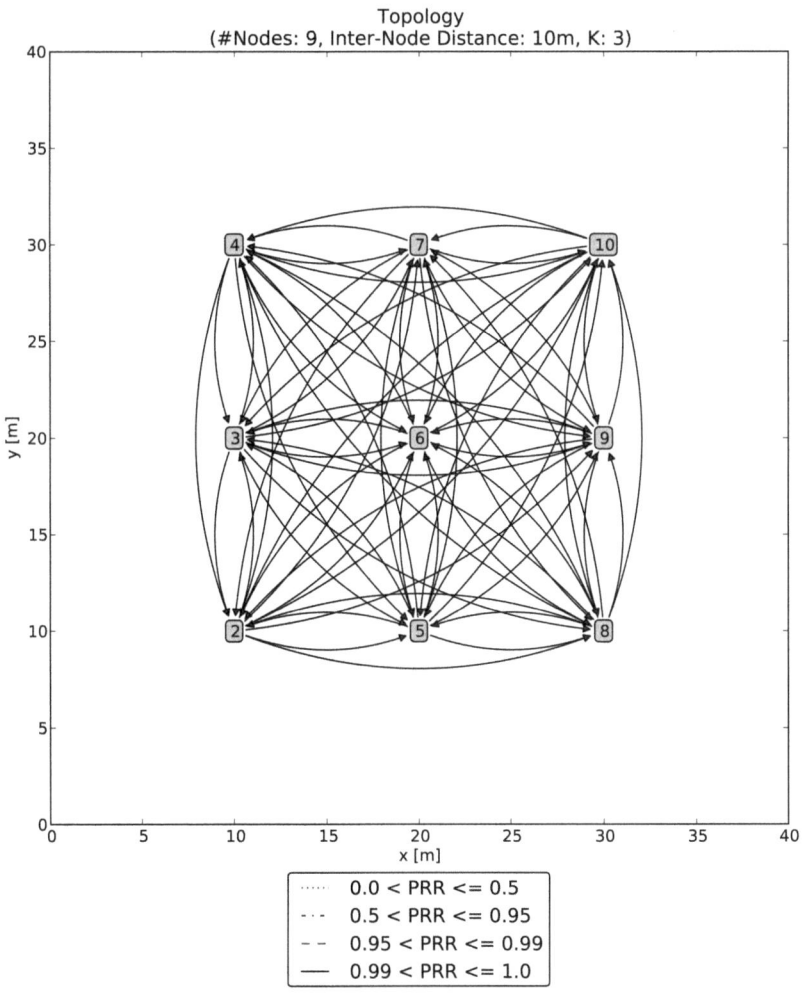

**Figure E.1: PRR Matrix (Distance 10m)**

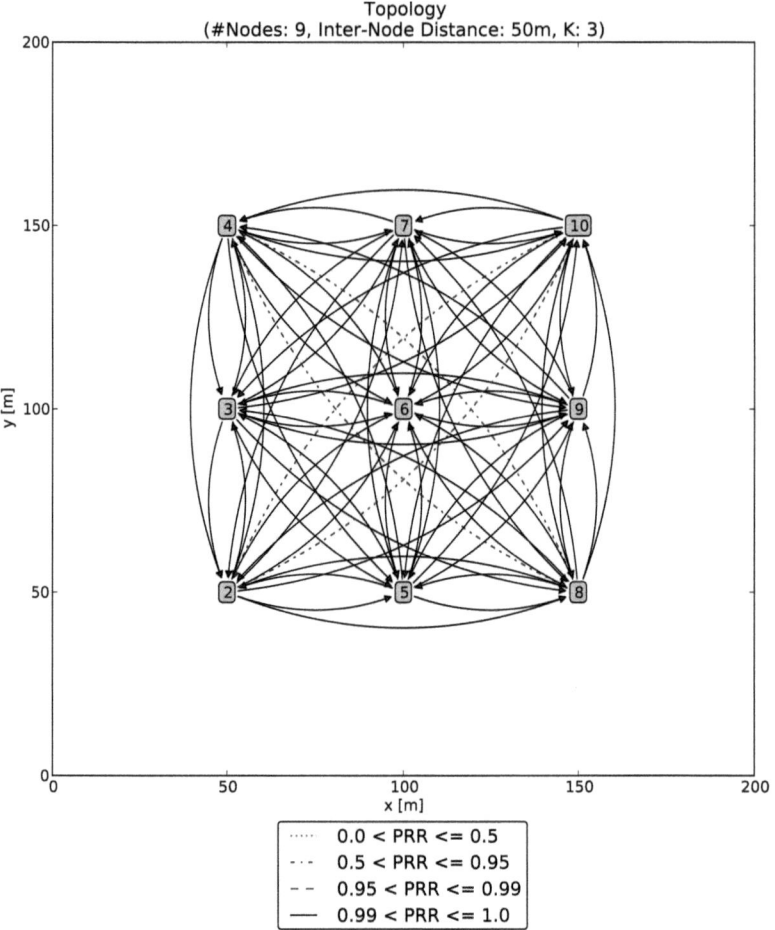

**Figure E.2: PRR Matrix (Distance 50m)**

# Simulated PRR Topologies (Grid-CPM)

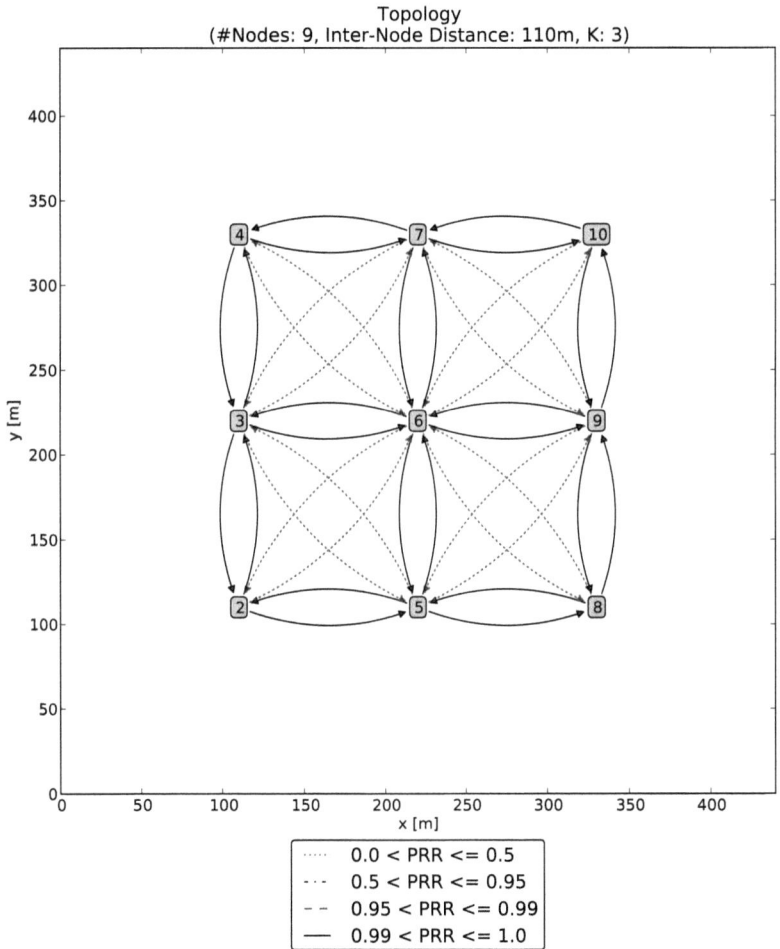

**Figure E.3: PRR Matrix (Distance 110m)**

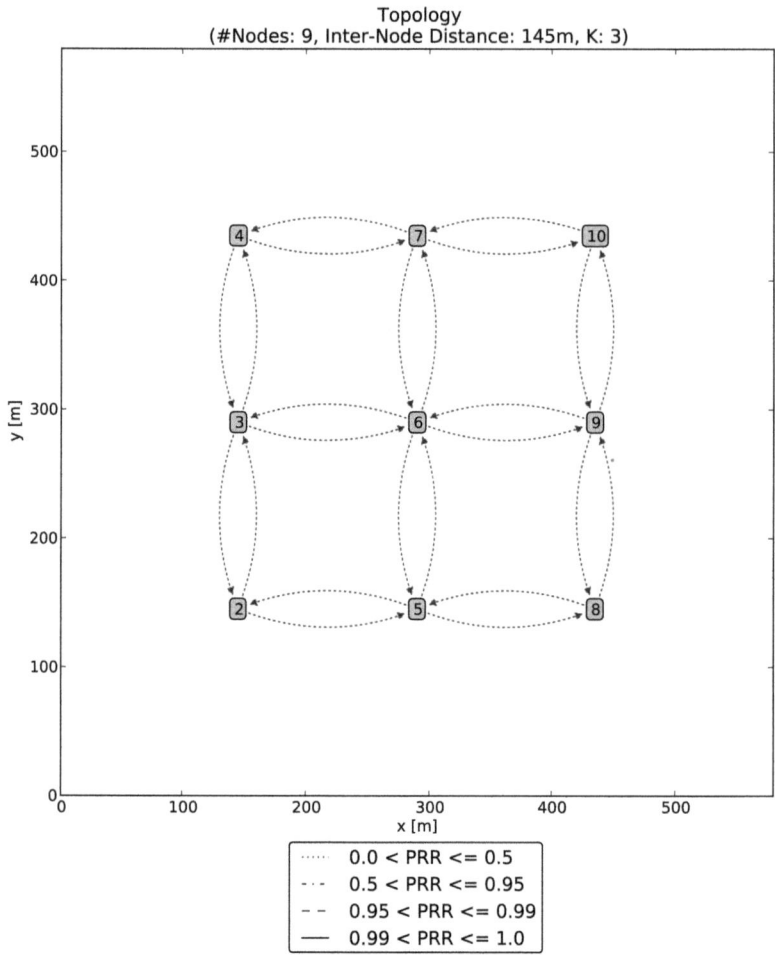

**Figure E.4: PRR Matrix (Distance 145m)**

# F Radio Models

## F.1 Signal Attenuation Model

The signal attenuation model used in the simulations is a regular log-distance model (as e.g. presented in [Rap09]) with typical parameters for IEEE 802.15.4 channels, confirm equation F.1.

$$S_{RX} = S_{TX} - (PL_0 + 10 * \gamma * \log_{10}(d)) \tag{F.1}$$

where

$S_{TX}$ Transmission Power
$S_{RX}$ Received Signal Strength
$PL_0$ Pathloss at 1 m distance
$\gamma$ Propagation coefficient
$d$ Distance in [m].

For the parameter $PL_0$ a value of 50 dBm is used, while for $\gamma$ the free-space value of 2 is used.

## F.2 Packet Reception Ratio Model

The Packet Reception Ratio (PRR) in TOSSIM is determined according to equation F.2 when using the CPM model [LCL07; Tin12].

$$PRR = (1 - 0.5 * \mathrm{erfc}(\frac{\beta_1 * \frac{S_{RX}}{N} - \beta_2}{\sqrt{2}}))^{23*2} \tag{F.2}$$

where

$S_{RX}$ Signal
$N$ Noise
$PRR$ Packet Reception Ratio.

The parameter $\beta_1$ is set to the value 0.9794 and $\beta_2$ is set to 2.3851 [Tin12].
A plot of the PRR-SNR model is shown in figure F.1.

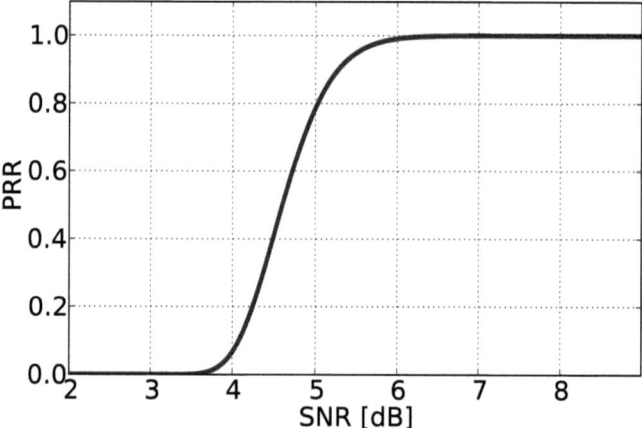

**Figure F.1: PRR-SNR Model**

# G Approximating Step-Wise Linear Model for the Time to Consistency

The evaluation of the resulting Cumulative Distribution Functions (cdfs) reveals that in most cases the cdf for the time to consistency can be modeled by a piecewise linear equation. In the sections G.1 and G.2 equations for such models are given.

## G.1 Approximating Linear Model for Line Scenarios

The approximating linear model for the line scenarios is given in equation G.1.

$$P(T \leq t) = \begin{cases} 0 & , t < 0 \\ \frac{1}{N} & , 0 \leq t < \frac{T_L}{2} \\ \frac{1}{N} + 0.8 \frac{N-1}{N} \frac{10}{d} \frac{10}{N}(t - \frac{T_L}{2}) & , \frac{T_L}{2} \leq t < t_{100\%} \\ 1 & , t \geq t_{100\%} \end{cases} \quad (G.1)$$

with $t_{100\%} = (1 - \frac{1}{N})\frac{1}{0.8}\frac{N}{N-1}\frac{d}{10}\frac{N}{10} + \frac{T_L}{2}$, where $N$ is the total number of nodes and $d$ is the inter-node distance.

## G.2 Approximating Linear Model for Grid Scenarios

In equation G.2 the approximating linear model for the grid scenarios is given.

$$P(T \leq t) = \begin{cases} 0 & , t < 0 \\ \frac{1}{N} & , 0 \leq t < \frac{T_L}{2} \\ \frac{1}{N} + 0.8 \frac{N-1}{N} \frac{10}{d} \frac{7.5}{\sqrt{N}}(t - \frac{T_L}{2}) & , \frac{T_L}{2} \leq t < t_{100\%} \\ 1 & , t \geq t_{100\%} \end{cases} \quad (G.2)$$

with $t_{100\%} = (1 - \frac{1}{N})\frac{1}{0.8}\frac{N}{N-1}\frac{d}{10}\frac{\sqrt{n}}{7.5} + \frac{T_L}{2}$.

# Bibliography

[AG07]  E. Avilés-López and J. A. García-Macías. "Providing Service-Oriented Abstractions for the Wireless Sensor Grid". In: *Advances in Grid and Pervasive Computing (Lecture Notes in Computer Science, Volume 4459/2007)*. ISSN: 0302-9743, ISBN: 978-3-540-72359-2. Springer Berlin / Heidelberg, 2007, pp. 710–715.

[Amb]  Ambient Systems. Available at: http://www.ambient-systems.net/. (Retrieved 02/17/2009).

[Amb13]  Ambient Systems. *SmartPoints 3000 Technology Overview White Paper*. Available at: http://www.ambient-systems.net/en/products/download_product_series_3000_white_paper.html. 2013. (Retrieved 07/15/2013).

[Amt04]  Amtsblatt der Europäischen Union. *Verordnung (EG) Nr. 852/2004 des Europäischen Parlaments und des Rates*. Amtsblatt der Europäischen Union. Kapitel IV, Paragraph 7. Apr. 2004.

[Arc]  Arch Rock. Available at: http://www.archrock.com/. (Retrieved 02/17/2009).

[Arc07]  Arch Rock. *Arch Rock IP/6LoWPAN Overview: An IPv6 Network Stack for Wireless Sensor Networks*. Tech. rep. Arch Rock, 2007.

[Atm13]  Atmel. *Atmel Zigbit Amp Datasheet*. Available at: http://www.atmel.com/Images/doc8228.pdf. 2013. (Retrieved 07/15/2013).

[BB65]  S. P. Burg and E. A. Burg. "Relationship between ethylene production and ripening in bananas". In: *Botanical Gazette* (1965), pp. 200–204.

[BBP10]  A. Brandt, J. Buron, and G. Porcu. *Home Automation Routing Requirements in Low-Power and Lossy Networks*. RFC 5826 (Informational). Apr. 2010. URL: http://www.ietf.org/rfc/rfc5826.txt.

[Bec+06a]  M. Becker, R. Jedermann, A. Sklorz, C. Behrens, D. Westphal, A. Timm-Giel, C. Görg, W. Lang, and R. Laur. *Sensoren und RFIDs für selbststeuernde logistische Objekte und deren mobile Vernetzung*. Berichtskolloquium des SFB 637. 2006.

[Bec+06b]  M. Becker, G. Sayyed, A. Timm-Giel, and C. Görg. *Analysis of Mobile Agents in Logistical Environments*. 17. Treffen der Informationstechnischen Gesellschaft, Fachgruppe 5.2.4. - Mobilität in IP-basierten Netzen zum Thema 'Communication Applications for Logistics: Maut, Telematics and More'. 2006.

[Bec+06c]  M. Becker, G. Sayyed, B.-L. Wenning, and C. Görg. "Analysis of Mobile Agents considering the Fan Out - Mobile Agents for Autonomous Logistics". In: *Proceedings of 2006 IEEE International Conference on Service Operations and Logistics, and Informatics. Shanghai, China*. 2006, pp. 511–515.

[Bec+06d]  M. Becker, B.-L. Wenning, C. Görg, J. D. Gehrke, M. Lorenz, and O. Herzog. "Agent-Based and Discrete Event Simulation of Autonomous Logistic Processes". In: *20th European Conference on Modelling and Simulation, Bonn, Sankt Augustin*. Ed. by W. Borutzky, A. Orsoni, and R. Zobel. May 2006, pp. 566–571. URL: http://www.scs-europe.net/services/ecms2006/ecms2006%20pdf/110-abs.pdf.

[Bec+06e]  M. Becker, B.-L. Wenning, A. Timm-Giel, and C. Görg. "Usage of Mobile Radio Services in Future Logistic Applications". In: *Mobilfunk Technologien und Anwendungen. 11. ITG-Fachtagung, 17. - 18. Mai 2006, Osnabrück*. 2006, pp. 137–141.

[Bec+07a]  M. Becker, K. Kuladinithi, A. Timm-Giel, and C. Görg. "Understanding Autonomous Cooperation in Logistics - The Impact on Management, Information and Communication and Material Flow. Springer". In: ed. by M. Hülsmann and K. Windt. Springer, 2007. Chap. Historical Development of the Idea of Self-Organization in Information and Communication Technology.

[Bec+07b]  M. Becker, G. Singh, B.-L. Wenning, and C. Görg. "On Mobile Agents for Autonomous Logistics: An Analysis of Mobile Agents considering the Fan Out and sundry Strategies". In: *International Journal of Services Operations and Logistics* 2.2 (2007).

[Bec+07c]  M. Becker, T. L. Weerawardane, X. Li, and C. Görg. "Extending OPNET Modeler with External Pseudo Random Number Generators and Statistical Evaluation by the Limited Relative Error Algorithm (Chapter 12)". In: *COST Action 285: Modeling and Simulation Tools for Research in Emerging Multi-Service Telecommunications*. University of Surrey, Guildford, UK, Mar. 2007.

[Bec+08a] M. Becker, R. Jedermann, A. Timm-Giel, and C. Görg. "Sensing and Communication Services for Food Transport Logistics". In: *CEWIT 2008*. Stony Brook, NY, USA, Oct. 2008, p. 21. URL: http://www.cewit.org/conference/docs/Proceedings2008_10-10-08.pdf.

[Bec+08b] M. Becker, A. Timm-Giel, K. Murray, C. Lynch, C. Görg, and D. Pesch. "Comparative Simulations of WSN". In: *ICT Mobile Summit 2008*. Ed. by P. Cunningham and M. Cunningham. Stockholm: Paul Cunningham and Miriam Cunningham (Eds) IIMC International Information Management Corporation, June 2008. ISBN: ISBN 978-1-905824-3.

[Bec+09] M. Becker, S. Yuan, R. Jedermann, A. Timm-Giel, W. Lang, and C. Görg. "Challenges of Applying Wireless Sensor Networks in Logistics". In: *CEWIT 2009. Wireless and IT driving Healthcare, Energy and Infrastructure Transformation*. 2009.

[Bec+10a] M. Becker, A. Timm-Giel, S. Das, and C. Görg. "TOSSDR: TinyOS on any Physical Layer implemented in Software Defined Radio". In: *7th European Conference on Wireless Sensor Networks 2010, EWSN 2010. Posters and Demos*. Ed. by S. Marusic, P. Pinto, J. Sa Silva, F. Boavida, B. Krishnamachari, and J. Pereira. Demo. University of Coimbra, 2010, pp. 21–23.

[Bec+10b] M. Becker, B.-L. Wenning, C. Görg, R. Jedermann, and A. Timm-Giel. "Logistic applications with Wireless Sensor Networks". In: *HotEmNets 2010*. 2010.

[Bec+11] M. Becker, T. Pötsch, K. Kuladinithi, and C. Görg. "Deployment of CoAP in Transport Logistics". In: *36th IEEE Conference on Local Computer Networks (LCN 2011)*. (Demonstration) Bonn, Germany, Oct. 2011.

[Bec+13] M. Becker, K. Li, K. Kuladinithi, and T. Pötsch. *Transport of CoAP over SMS, USSD and GPRS*. Internet-Draft: draft-becker-core-coap-sms-gprs-03 (Work in Progress). Feb. 2013.

[Bec10] M. Becker. *Test des transparenten Datenkanals und Netzabdeckung bei Rungis Express*. Tech. rep. Bremen: Communication Networks, Universität Bremen, 2010.

[Bec13] M. Becker. *A cheatsheet for the Constrained Application Protocol (CoAP)*. Available at: https://github.com/markushx/coap-cheatsheet. Jan. 2013. (Retrieved 07/13/2013).

[Beh+06a]  C. Behrens, M. Becker, J. D. Gehrke, R. Jedermann, C. Görg, O. Herzog, W. Lang, and R. Laur. "Ein Multi-Agentensystem für Selbststeuerung in der Transportlogistik". In: *VDE-Kongress 2006, Oktober 2006, Aachen, Germany. Fachtagungsberichte VDE Kongress 2006 - Innovations for Europe, 23.-25. Okt 2006, Aachen.* 2006, pp. 29–34.

[Beh+06b]  C. Behrens, M. Becker, J. D. Gehrke, D. Peters, and R. Laur. "Wireless Sensor Networks as an Enabler for Cooperating Logistic Processes". In: *Proceedings of the ACM Workshop on Real-World Wireless Sensor Networks 2006 (REALWSN'06)*. ACM, June 2006, pp. 85–86. URL: http://www.tzi.de/~jgehrke/publikationen.html.

[Beh+06c]  C. Behrens, M. Becker, J. D. Gehrke, D. Peters, and R. Laur. "Wireless Sensor Networks as an Enabler for Cooperating Logistic Processes". In: *REALWSN06 ACM Workshop on Real-World Wireless Sensor Networks.* 2006, pp. 85–86.

[Beh09]  C. Behrens. *Kooperatives Energiemanagement in ressourcenbeschränkten Systemen.* Verl. Dr. Hut, 2009.

[Ber10]  O. Bergmann. *libcoap: C-Implementation of CoAP.* Available at http://libcoap.sourceforge.net/. 2010. (Retrieved 01/23/2013).

[BG06]  M. Becker and C. Görg. *Embedded Internet Telephony - Access Point, Bluetooth, Asterisk.* 19. Treffen der Informationstechnischen Gesellschaft, Fachgruppe 5.2.4 zum Thema - VoIP over Wireless. 2006.

[BJ08]  M. Becker and R. Jedermann. *T4 Wireless Sensor Network Propagation Analysis.* Tech. rep. SFB637-T4-2008-1. Bremen: Communication Networks, University Bremen, 2008.

[BJ09]  M. Becker and R. Jedermann. *Message Format Specification.* Tech. rep. SFB637-T4-2009-6. Bremen: Communication Networks, Universität Bremen, 2009.

[BKG10]  M. Becker, K. Kuladinithi, and C. Görg. "IETF Standards for Wireless Sensor Networks and Deployments in Logistic Applications". In: *5th International Conference on Information and Automation for Sustainability.* Colombo, Sri Lanka, Dec. 2010.

[BKG11a]  M. Becker, K. Kuladinithi, and C. Görg. "The Intelligent Container - Wireless Sensor Networks in Logistics". In: *19th ComNets Workshop.* Aachen, Germany, Mar. 2011.

# Bibliography

[BKG11b]  M. Becker, K. Kuladinithi, and C. Görg. *Tutorial on Wireless Sensor Networks, Institute of Engineers, Sri Lanka, Colombo*. Tutorial on Wireless Sensor Networks. 2011.

[Bou+08]  B. Bougard, F. Catthoor, D. C. Daly, A. Chandrakasan, and W. Dehaene. "Energy efficiency of the IEEE 802.15. 4 standard in dense wireless microsensor networks: Modeling and improvement perspectives". In: *Design, Automation, and Test in Europe*. Springer. 2008, pp. 221–234.

[BPK13]  M. Becker, T. Pötsch, and K. Kuladinithi. *Scenarios for CoAP on non-UDP Transports*. Internet-Draft: draft-becker-core-transport-scenarios-00 (Work in Progress). July 2013.

[BR00]  C. Bettstetter and C. Renner. "A Comparison of Service Discovery Protocols and Implementation of the Service Location Protocol." In: *Proc of EUNICE 2000, Sixth EUNICE Open European Summer School*. Twente, Netherlands, Sept. 2000.

[BT05]  M. Becker and A. Timm-Giel. "Selbststeuerung in der Transportlogistik: Modellierung der mobilen Kommunikation". In: *Industrie Management 5/2005* (2005), pp. 71–74. URL: http://www.comnets.uni-bremen.de/~mab/publications/2005/IM-B3-final-bscw.pdf.

[BTG06]  M. Becker, A. Timm-Giel, and C. Görg. *Vergleich von Simulationswerkzeugen für die Simulation von drahtlosen Sensornetzen*. Treffen der ITG FG 5.2.1. 2006.

[BTG07]  M. Becker, A. Timm-Giel, and C. Görg. "Understanding Autonomous Cooperation in Logistics - The Impact on Management, Information and Communication and Material Flow. Springer". In: ed. by M. Hülsmann and K. Windt. Springer, 2007. Chap. Self-Organization Concepts for the Information- and Communication Layer.

[BTG08]  M. Becker, A. Timm-Giel, and C. Görg. *Heterogeneous Network Access of the Intelligent Container*. LDIW '08, LogDynamics International Workshop 2008. 2008.

[BWG06]  M. Becker, B.-L. Wenning, and C. Görg. "Integrated Simulation of Communication Networks and Logistical Networks - Using Object Oriented Programming Language Features to Enhance Modelling". In: *Modeling and Simulation Tools for Emerging Telecommunications Networks - Needs, Trends, Challenges, Solutions. Springer. Ed.*

|         |                                                                                                                                                                                                                                                                                                            |
| ------- | ---------------------------------------------------------------------------------------------------------------------------------------------------------------------------------------------------------------------------------------------------------------------------------------------------------- |
|         | by N. Ince and E. Topuz. 2006, pp. 279–287. URL: http://www.comnets.uni-bremen.de/~mab/publications/2005/COST285-final-bscw.pdf.                                                                                                                                                                            |
| [CK13a] | S. Cheshire and M. Krochmal. *DNS-Based Service Discovery*. RFC 6763 (Proposed Standard). Feb. 2013. URL: http://www.ietf.org/rfc/rfc6763.txt.                                                                                                                                                              |
| [CK13b] | S. Cheshire and M. Krochmal. *Multicast DNS*. RFC 6762 (Proposed Standard). Feb. 2013. URL: http://www.ietf.org/rfc/rfc6762.txt.                                                                                                                                                                            |
| [Col]   | Collaborative Research Centre 637. *Collaborative Research Centre 637: Autonomous Cooperating Logistic Processes – A Paradigm Shift and its Limitations*. Available at: http://www.sfb637.uni-bremen.de. (Retrieved 05/06/2010).                                                                            |
| [Cro]   | Crossbow. Available at: http://www.xbow.com. (Retrieved 02/17/2009).                                                                                                                                                                                                                                       |
| [Ded12] | J. Dede. *Evaluation of the Trickle Algorithm for Service Discovery in Wireless Sensor Networks*. Studienarbeit. 2012.                                                                                                                                                                                     |
| [Del+05]| F. C. Delicato, P. F. Pires, L. Rust, L. Pirmez, and J. F. de Rezende. "Reflective middleware for wireless sensor networks". In: *Proceedings of the 2005 ACM Symposium on Applied Computing*. Ed. by L. M. Liebrock. ACM, New York, 1155-1159, 2005. DOI: http://doi.acm.org/10.1145/1066677.1066937. (Retrieved 03/30/2010). |
| [DGV04] | A. Dunkels, B. Gronvall, and T. Voigt. "Contiki - A Lightweight and Flexible Operating System for Tiny Networked Sensors". In: *LCN '04: Proceedings of the 29th Annual IEEE International Conference on Local Computer Networks*. Washington, DC, USA: IEEE Computer Society, 2004, pp. 455–462. ISBN: 0-7695-2260-2. DOI: http://dx.doi.org/10.1109/LCN.2004.38. (Retrieved 02/17/2009). |
| [DH95]  | S. Deering and R. Hinden. *Internet Protocol, Version 6 (IPv6) Specification*. RFC 1883 (Proposed Standard). Obsoleted by RFC 2460. Dec. 1995. URL: http://www.ietf.org/rfc/rfc1883.txt.                                                                                                                    |
| [DH98]  | S. Deering and R. Hinden. *Internet Protocol, Version 6 (IPv6) Specification*. RFC 2460 (Draft Standard). Updated by RFCs 5095, 5722, 5871, 6437, 6564. Dec. 1998. URL: http://www.ietf.org/rfc/rfc2460.txt.                                                                                                |

[Doh+09]   M. Dohler, T. Watteyne, T. Winter, and D. Barthel. *Routing Requirements for Urban Low-Power and Lossy Networks*. RFC 5548 (Informational). May 2009. URL: http://www.ietf.org/rfc/rfc5548.txt.

[DS08]   K. Daschkovska and B. Scholz-Reiter. "Electronic seals for efficient container logistics". In: *Dynamics in Logistics*. Springer, 2008, pp. 305–312.

[Dur+08]   M. Durvy, J. Abeillé, P. Wetterwald, C. O'Flynn, B. Leverett, E. Gnoske, M. Vidales, G. Mulligan, N. Tsiftes, N. Finne, and A. Dunkels. "Making sensor networks IPv6 ready". In: *The Sixth ACM Conference on Networked Embedded Sensor Systems (ACM SenSys 2008)*. Raleigh, North Carolina, USA., Nov. 2008.

[Est+02]   D. Estrin, D. Culler, K. Pister, and G. Sukhatme. "Connecting the physical world with pervasive networks". In: *Pervasive Computing, IEEE* 1.1 (2002), pp. 59–69.

[Eur99]   European Committee for Electrotechnical Standardization. *EN 12830:1999 Temperature recorders for the transport, storage and distribution of chilled, frozen, deep-frozen/quick-frozen food and ice cream - Tests, performance, suitability*. 1999.

[FHK04]   C. Frank, V. Handziski, and H. Karl. *Service Discovery in Wireless Sensor Networks*. Tech. rep. TKN-04-006. Available at: http://www.tkn.tu-berlin.de/publications/papers/handziski_Serv_Discov_In_WSN1.pdf. TKN, Mar. 2004. (Retrieved 03/30/2010).

[For+07]   I. Forkel, M. Schmocker, L. Lazin, M. Becker, F. Debus, and F. Winnewisser. "Cell-Specific Optimized Parameterization of Compressed Mode Operation and Inter-System Handovers in UMTS/GSM Overlay Networks". In: *13th European Wireless Conference 2007. Paris, France*. 2007. URL: http://www.ew2007.org/papers/1569014932.pdf.

[Geh+06]   J. D. Gehrke, B.-L. Wenning, M. Lorenz, and M. Becker. "Integration of two Approaches for Simulation of Autonomous Logistic Processes". In: *Simulation in Produktion und Logistik 2006. Tagungsband zur 12. Fachtagung*. Ed. by S. Wenzel. Kassel, Germany: SCS Publishing House, Sept. 2006, pp. 133–142. URL: http://www.tzi.de/~jgehrke/publikationen.html.

[Gol+03] F. Golatowski, J. Blumenthal, M. Handy, M. Haase, H. Burchardt, and D. Timmermann. "Service-oriented software architecture for sensor networks". In: *Proceedings of the International Workshop on Mobile Computing*. 2003, pp. 93–98.

[Gut+99] E. Guttman, C. Perkins, J. Veizades, and M. Day. *Service Location Protocol, Version 2*. RFC 2608 (Proposed Standard). Updated by RFC 3224. June 1999. URL: http://www.ietf.org/rfc/rfc2608.txt.

[Han+05] C.-C. Han, R. Kumar, R. Shea, E. Kohler, and M. Srivastava. "A dynamic operating system for sensor nodes". In: *MobiSys '05: Proceedings of the 3rd international conference on Mobile systems, applications, and services*. Seattle, Washington: ACM, 2005, pp. 163–176. ISBN: 1-931971-31-5. DOI: http://doi.acm.org/10.1145/1067170.1067188. (Retrieved 02/17/2009).

[Har+07] S. Harte, B. O'Flynn, R. Martinez-Catala, and E. Popovici. "Design and implementation of a miniaturised, low power wireless sensor node". In: *18th European Conference on Circuit Theory and Design, 2007. ECCTD 2007*. 978-1-4244-1341-6. Aug. 2007, pp. 894–897. DOI: 10.1109/ECCTD.2007.4529741. (Retrieved 02/17/2009).

[Hau05] J.-H. Hauer. "Service discovery in wireless sensor networks using publish/subscribe middleware". MA thesis. Telecommunication Networks Group, Technische Universität Berlin, 2005.

[Hau09] J.-H. Hauer. *TKN15.4: An IEEE 802.15.4 MAC Implementation for TinyOS 2*. Tech. rep. TKN-08-00. Telecommunication Networks Group, Technical University Berlin, Mar. 2009.

[HC08] J. W. Hui and D. E. Culler. "IP is dead, long live IP for wireless sensor networks". In: *SenSys '08: Proceedings of the 6th ACM conference on Embedded network sensor systems*. Raleigh, NC, USA: ACM, 2008, pp. 15–28. ISBN: 978-1-59593-990-6. DOI: http://doi.acm.org/10.1145/1460412.1460415. (Retrieved 05/03/2009).

[Hew04] Hewlett-Packard Company. *HP Photosmart networking guide*. Aug. 2004.

[HS08] M. Harvan and J. Schönwälder. "TinyOS Motes on the Internet: IPv6 over 802.15.4". In: *PIK - Praxis der Informationsverarbeitung und Kommunikation* 31.4 (2008), pp. 244–251. DOI: 10.1515/piko.2008.0042. (Retrieved 08/03/2009).

# Bibliography

[HT11] J. Hui and P. Thubert. *Compression Format for IPv6 Datagrams over IEEE 802.15.4-Based Networks*. RFC 6282 (Proposed Standard). Sept. 2011. URL: http://www.ietf.org/rfc/rfc6282.txt.

[Hui+12] J. Hui, J. Vasseur, D. Culler, and V. Manral. *An IPv6 Routing Header for Source Routes with the Routing Protocol for Low-Power and Lossy Networks (RPL)*. RFC 6554 (Proposed Standard). Mar. 2012. URL: http://www.ietf.org/rfc/rfc6554.txt.

[Hus09] G. Huston. *IPv4 Address Report, daily generated*. Available at: http://www.potaroo.net/tools/ipv4/index.html. Apr. 2009. (Retrieved 04/26/2009).

[HV12] J. Hui and J. Vasseur. *The Routing Protocol for Low-Power and Lossy Networks (RPL) Option for Carrying RPL Information in Data-Plane Datagrams*. RFC 6553 (Proposed Standard). Mar. 2012. URL: http://www.ietf.org/rfc/rfc6553.txt.

[IEE03] IEEE Computer Society. "Part 15.4: Wireless Medium Access Control (MAC) and Physical Layer (PHY) Specifications for Low-Rate Wireless Personal Area Networks (WPANs)". In: *IEEE Standard for Information technology - Telecommunications and information exchange between systems - Local and metropolitan area networks - Specific requirements*. IEEE Std 802.15.4-2003. May 2003.

[IEE06] IEEE Computer Society. "Part 15.4: Wireless Medium Access Control (MAC) and Physical Layer (PHY) Specifications for Low-Rate Wireless Personal Area Networks (WPANs)". In: *IEEE Standard for Information technology- Telecommunications and information exchange between systems - Local and metropolitan area networks - Specific requirements*. IEEE Std 802.15.4-2006. Sept. 2006.

[Int] Internet Engineering Task Force. *IPv6 over Low power WPAN (6LoWPAN) – Working Group charter*. Available at: http://www.ietf.org/html.charters/6lowpan-charter.html. (Retrieved 04/26/2009).

[Int13] Intel. *Imote2 Datasheet*. Available at: http://bullseye.xbow.com:81/Products/Product_pdf_files/Wireless_pdf/Imote2_Datasheet.pdf. 2013. (Retrieved 07/15/2013).

[IPS] IPSO Alliance. *IPSO Alliance*. Available at: http://www.ipso-alliance.org. (Retrieved 04/27/2010).

[ISO07]   ISO. *ISO/IEC DIS 29341: Information technology – UPnP Device Architecture 1.0*. Tech. rep. International Organization for Standardization, 2007.

[JBY08]   R. Jedermann, M. Becker, and S. Yuan. *Report Sensor Test at Dole Hamburg / Stelle, October and November 2008, Part I Temperature and Humidity measurements*. Tech. rep. SFB637-T4-2008-4. Bremen: Institute for Microsensors, -actuators and -systems, University Bremen, 2008.

[Jed+07]   R. Jedermann, J. D. Gehrke, M. Becker, C. Behrens, E. M. Kluge, O. Herzog, and W. Lang. "Understanding Autonomous Cooperation & Control in Logistics. The Impact on Management, Information and Communication and Material Flow". In: ed. by M. Hülsmann and K. Windt. Berlin: Springer, 2007. Chap. Transport Scenario for the Intelligent Container, pp. 393–404. DOI: 10.1007/978-3-540-47450-0_25.

[Jed+08a]   R. Jedermann, M. Becker, S. Yuan, and L. Chen. *Report Sensor Test at Dole Hamburg / Stelle October and November 2008, Part II Radio signal Attenuation by bananas*. Tech. rep. SFB637-T4-2008-5. Bremen: Institute for Microsensors, -actuators and -systems, University Bremen, 2008.

[Jed+08b]   R. Jedermann, K. Stein, M. Becker, and W. Lang. "UHF-RFID in the Food Chain - From Identification to Smart Labels". In: *Coldchain Manangement. 3rd International Workshop*. Ed. by J. Kreyenschmidt. Bonn, 2008.

[Jed+10a]   R. Jedermann, M. Becker, C. Görg, and W. Lang. "Field Test of the Intelligent Container". In: *7th European Conference on Wireless Sensor Networks EWSN2010. Posters and Demos*. Coimbra: University of Coimbra, 2010, pp. 11–12.

[Jed+10b]   R. Jedermann, W. Lang, M. Becker, X. Wang, and A. Jabbari. *Der Intelligente Container*. Schadenverhütungstagung 2010 - Temperaturgeführte Transporte. 2010.

[Jed+11]   R. Jedermann, M. Becker, C. Görg, and W. Lang. "Testing network protocols and signal attenuation in packed food transports". In: *International Journal of Sensor Networks (IJSNet)* 9.3 (2011), pp. 170–181.

[Jed09]   R. Jedermann. *Autonome Sensorsysteme in der Transport- und Lebensmittellogistik*. Verlag Dr. Hut, 2009.

[Jia+09] X. Jiang, S. Dawson-Haggerty, P. Dutta, and D. Culler. "Design and Implementation of a High-Fidelity AC Metering Network". In: *Proceedings of the Eighth International Conference on Information Processing in Sensor Networks (IPSN'09) Track on Sensor Platforms, Tools, and Design Methods (SPOTS '09)*. Apr. 2009.

[KH08] A. Köpke and J.-H. Hauer. *IEEE 802.15.4 Symbol Rate Timer for TelosB*. Tech. rep. TKN-08-006. Telecommunication Networks Group, Technische Universität Berlin, May 2008.

[Kim+07] K. Kim, S. Yoo, H. Lee, S. D. Park, and J. Lee. *Simple Service Location Protocol (SSLP) for 6LoWPAN*. Tech. rep. Available at: http://tools.ietf.org/html/draft-daniel-6lowpan-sslp-01. IETF, June 2007.

[Kim+12] E. Kim, D. Kaspar, C. Gomez, and C. Bormann. *Problem Statement and Requirements for IPv6 over Low-Power Wireless Personal Area Network (6LoWPAN) Routing*. RFC 6606 (Informational). May 2012. URL: http://www.ietf.org/rfc/rfc6606.txt.

[KM58] E. L. Kaplan and P. Meier. "Nonparametric estimation from incomplete observations". In: *Journal of the American statistical association* 53.282 (1958), pp. 457–481.

[KMS07] N. Kushalnagar, G. Montenegro, and C. Schumacher. *IPv6 over Low-Power Wireless Personal Area Networks (6LoWPANs): Overview, Assumptions, Problem Statement, and Goals*. RFC 4919 (Informational). Aug. 2007. URL: http://www.ietf.org/rfc/rfc4919.txt.

[KP80] J. D. Kalbfleisch and R. L. Prentice. *The statistical analysis of failure time data*. Wiley series in probability and mathematical statistics. XI, 321 S : graph. Darst. New York, NY [u.a.]: Wiley, 1980. ISBN: 0471055190.

[Kre03] C. Kredler. *Materialien zu Stochastik 1, Einführung in die Wahrscheinlichkeitsrechnung und Statistik*. Available at: http://www.ma.tum.de/foswiki/pub/Studium/ChristianKredler/Stoch1.pdf. 2003. (Retrieved 12/02/2012).

[Kul+05] K. Kuladinithi, M. Becker, C. Görg, and S. R. Das. *Radio Disjoint Multi-Path Routing in MANET*. CEWIT (Center of Excellence in Wireless and Information Technology) 2005 Conference, Stony Brook. 2005. URL: http://www.comnets.uni-bremen.de/~koo/2005cewit.ppt.

[Kul+11] K. Kuladinithi, O. Bergmann, T. Pötsch, M. Becker, and C. Görg. "Implementation of CoAP and its Application in Transport Logistics". In: *Extending the Internet to Low power and Lossy Networks (IP+SN 2011)*. Chicago, USA, Apr. 2011.

[LCL07] H. Lee, A. Cerpa, and P. Levis. "Improving wireless simulation through noise modeling". In: *IPSN '07: Proceedings of the 6th international conference on Information processing in sensor networks*. ACM Press, 2007, pp. 21–30.

[Lev+03] P. Levis, N. Lee, M. Welsh, and D. Culler. "TOSSIM: Accurate and scalable simulation of entire TinyOS applications". In: *Proceedings of the 1st international conference on Embedded networked sensor systems*. ACM. 2003, pp. 126–137. ISBN: 1581137079.

[Lev+04] P. Levis, N. Patel, D. Culler, and S. Shenker. "Trickle: a self-regulating algorithm for code propagation and maintenance in wireless sensor networks". In: *NSDI'04: Proceedings of the 1st conference on Symposium on Networked Systems Design and Implementation*. Berkeley, CA, USA: USENIX Association, 2004, pp. 2–2.

[Lev+05] P. Levis, S. Madden, J. Polastre, R. Szewczyk, K. Whitehouse, A. Woo, D. Gay, J. Hill, M. Welsh, E. Brewer, et al. "TinyOS: An operating system for sensor networks". In: *Ambient intelligence*. Springer, 2005, pp. 115–148.

[Lev+11] P. Levis, T. Clausen, J. Hui, O. Gnawali, and J. Ko. *The Trickle Algorithm*. RFC 6206 (Proposed Standard). Mar. 2011. URL: http://www.ietf.org/rfc/rfc6206.txt.

[LK01] H. Lim and C. Kim. "Flooding in wireless ad hoc networks". In: *Computer Communications* 24.3-4 (2001), pp. 353–363. ISSN: 0140-3664. DOI: 10.1016/S0140-3664(00)00233-4.

[Lor+11] S. Loreto, P. Saint-Andre, S. Salsano, and G. Wilkins. *Known Issues and Best Practices for the Use of Long Polling and Streaming in Bidirectional HTTP*. RFC 6202 (Informational). Apr. 2011. URL: http://www.ietf.org/rfc/rfc6202.txt.

[MEM13] MEMSIC. *TelosB Datasheet*. Available at: http://www.memsic.com/userfiles/files/Datasheets/WSN/telosb_datasheet.pdf. 2013. (Retrieved 07/15/2013).

# Bibliography

[ML08]  D. Moss and P. Levis. "BoX-MACs: Exploiting physical and link layer boundaries in low-power networking". In: *Computer Systems Laboratory Stanford University* (2008).

[Mon+07]  G. Montenegro, N. Kushalnagar, J. Hui, and D. Culler. *Transmission of IPv6 Packets over IEEE 802.15.4 Networks*. RFC 4944 (Proposed Standard). Updated by RFCs 6282, 6775. Sept. 2007. URL: http://www.ietf.org/rfc/rfc4944.txt.

[Öst+06]  Å. Östmark, P. Lindgren, A. Van Halteren, and L. Meppelink. "Service and device discovery of nodes in a wireless sensor network". In: (2006).

[PBG13]  T. Pötsch, M. Becker, and C. Görg. "Continuous Integration for Wireless Sensor Network Operating Systems". In: *10th European Conference on Wireless Sensor Networks (EWSN), (Poster)*. Ghent, Belgium, Feb. 2013.

[Pis+09]  K. Pister, P. Thubert, S. Dwars, and T. Phinney. *Industrial Routing Requirements in Low-Power and Lossy Networks*. RFC 5673 (Informational). Oct. 2009. URL: http://www.ietf.org/rfc/rfc5673.txt.

[PL03]  R. Perrey and M. Lycett. "Service-oriented architecture". In: *Applications and the Internet Workshops, 2003. Proceedings. 2003 Symposium on*. IEEE. 2003, pp. 116–119.

[Poe06]  L. Poettering. "Null Arbeit: Zeroconf-Netzwerktechniken unter Linux mit Avahi nutzen". In: *Linux Magazin* 3 (2006). Available at: http://www.linux-magazin.de/heft\_abo/ausgaben/2006/03/null\_arbeit, p. 64.

[Pos80]  J. Postel. *DoD standard Internet Protocol*. RFC 760. Obsoleted by RFC 791, updated by RFC 777. Jan. 1980. URL: http://www.ietf.org/rfc/rfc760.txt.

[PP08]  R. de Paz Alberola and D. Pesch. "AvroraZ: extending Avrora with an IEEE 802.15.4 compliant radio chip model". In: *Proceedings of the 3nd ACM workshop on Performance monitoring and measurement of heterogeneous wireless and wired networks*. PM2HW2N '08. Vancouver, British Columbia, Canada: ACM, 2008, pp. 43–50. ISBN: 978-1-60558-239-9. DOI: 10.1145/1454630.1454637.

[PSC05]   J. Polastre, R. Szewczyk, and D. Culler. "Telos: enabling ultra-low power wireless research". In: *Information Processing in Sensor Networks, 2005. IPSN 2005. Fourth International Symposium on.* IEEE. 2005, pp. 364–369.

[R D12]   R Development Core Team. *The R Manuals.* Available at: http://cran.r-project.org/manuals.html. June 2012. (Retrieved 01/31/2013).

[Rap09]   T. S. Rappaport. *Wireless communications: principles and practice.* 2. ed., 18. printing. Prentice Hall communications engineering and emerging technologies series. XXIII, 707 S. : Ill., graph. Darst. Upper Saddle River, NJ: Prentice Hall PTR, 2009. ISBN: 0130422320 and 9780130422323.

[RB08]   S. Rumley and M. Becker. "Pseudo Random Numbers Generators available as Web Services". In: *In Proceedings of 2008 International Symposium on Performance Evaluation of Computer and Telecommunication Systems (SPECTS).* Edinburgh, UK, June 2008.

[RE07]   A. Rezgui and M. Eltoweissy. "Service-Oriented Sensor-Actuator Networks [Ad Hoc and Sensor Networks]". In: *Communications Magazine, IEEE* 45.12 (Dec. 2007). ISSN: 0163-6804, pp. 92–100.

[Rei02]   H. Reichl. "The eGrain system-using fine electronic particles". In: *Fraunhofer magazine* 1 (2002).

[SB10]   Z. Shelby and C. Bormann. *6LoWPAN: The Wireless Embedded Internet.* Wiley, 2010. ISBN: 0470747994.

[Sena]   Sentilla (formerly Moteiv). Available at: http://www.sentilla.com. (Retrieved 02/17/2009).

[Senb]   Sensinode Ltd. Available at: http://www.sensinode.com/. (Retrieved 02/17/2009).

[She+13]   Z. Shelby, K. Hartke, C. Bormann, and B. Frank. *Constrained Application Protocol (CoAP).* Available at: http://tools.ietf.org/html/draft-ietf-core-coap-18. Internet Engineering Task Force, June 28, 2013. (Retrieved 07/13/2013).

[Sin+07]   G. Singh, B.-L. Wenning, M. Becker, A. Timm-Giel, and C. Görg. "Agent-based Clustering Approach to Transport Logistics". In: *2007 IEEE International Conference on Service Operations and Logistics, and Informatics, Philadelphia, USA.* 2007, pp. 466–471.

[Son+05]   H. Song, D. Kim, K. Lee, and J. Sung. "UPnP-Based Sensor Network Management Architecture". In: *Proc. of ICMU.* 2005.

# Bibliography

[Sun] Sun. *SunSpots*. Available at: http://www.sunspotworld.com. (Retrieved 02/17/2009).

[Sun07] Sun. *Jini Architectural Overview*. Available at: http://www.sun.com/jini/. 2007. (Retrieved 04/07/2012).

[TBG06] A. Timm-Giel, M. Becker, and C. Görg. *Wireless Sensor Networks in Wearable and Logistic Application, Joint e-SENSE*. MAGNET Beyond, DAIDALOS, and CRUISE Workshop during the IST Mobile Summit 2006 on Myconos (GR), June 8th. 2006.

[TBG11] P. Trenkamp, M. Becker, and C. Görg. "Wireless Sensor Network Platforms - Datasheets versus Measurements". In: *Sixth IEEE International Workshop on Practical Issues in Building Sensor Network Applications (SenseApp 2011)*. Bonn, Germany, Oct. 2011.

[Tex07] Texas Instruments. *CC2420 datasheet*. Available at: http://focus.ti.com/docs/prod/folders/print/cc2420.html. 2007. (Retrieved 05/10/2011).

[The12] T. Therneau. *The Comprehensive R Archive Network: Module Survival*. Available at: http://cran.r-project.org/web/packages/survival/index.html. Apr. 2012. (Retrieved 01/31/2013).

[Thu12] P. Thubert. *Objective Function Zero for the Routing Protocol for Low-Power and Lossy Networks (RPL)*. RFC 6552 (Proposed Standard). Mar. 2012. URL: http://www.ietf.org/rfc/rfc6552.txt.

[Tim+07a] A. Timm-Giel, M. Becker, B.-L. Wenning, C. Görg, C. Lynch, and D. Pesch. *Simulation and Modelling of Wireless Sensor Networks*. CRUISE/e-SENSE Workshop at VTC Dublin 2007. 2007.

[Tim+07b] A. Timm-Giel, K. Murray, M. Becker, C. Guo, R. Sokullu, and D. Marandin. "Application Framework and Network Architecture for Wireless Sensor Networks". In: *Globecom 2007*. Nov. 2007.

[Tin07] TinyOS. *TinyOS*. Available at: http://www.tinyos.net. 2007. (Retrieved 07/23/2007).

[Tin12] TinyOS Contributors. *TinyOS Source Code Repository*. Available at https://github.com/tinyos/tinyos-main. 2012. (Retrieved 07/13/2013).

[TZI] TZI. *6LoWPAN Wiki – IPv6 over Low-Power Wireless Personal Area Networks*. Available at: http://6lowpan.tzi.org/. (Retrieved 04/26/2009).

[US 10]   U.S. Energy Information Administration. "International Energy Outlook 2010". In: Report #: DOE/EIA-0484(2010. Table D1. Available at: http://www.eia.doe.gov/oiaf/ieo/. Office of Integrated Analysis and Forecasting, U.S. Department of Energy, Washington DC 20585, May 2010. Chap. 1, p. 207.

[Vas+12]  J. Vasseur, M. Kim, K. Pister, N. Dejean, and D. Barthel. *Routing Metrics Used for Path Calculation in Low-Power and Lossy Networks*. RFC 6551 (Proposed Standard). Mar. 2012. URL: http://www.ietf.org/rfc/rfc6551.txt.

[Vir13]   Virtenio. *Preon32 Datasheet*. Available at: http://www.virtenio.com/en/assets/downloads/datenblaetter/DS_Preon32_v14_2page\%20[EN].pdf. 2013. (Retrieved 07/15/2013).

[WBG05]   B.-L. Wenning, M. Becker, and C. Görg. "Applying Internet Routing to Logistics: The Border Gateway Protocol". In: *Operations Research Conference*. 2005.

[Wen+07]  B.-L. Wenning, H. Rekersbrink, M. Becker, A. Timm-Giel, C. Görg, and B. Scholz-Reiter. "Understanding Autonomous Cooperation & Control in Logistics - The Impact on Management, Information and Communication and Material Flow. Springer, Berlin". In: ed. by M. Hülsmann and K. Windt. Springer, 2007. Chap. Dynamic transport reference scenarios, pp. 337–350.

[Win+12]  T. Winter, P. Thubert, A. Brandt, J. Hui, R. Kelsey, P. Levis, K. Pister, R. Struik, J. Vasseur, and R. Alexander. *RPL: IPv6 Routing Protocol for Low-Power and Lossy Networks*. RFC 6550 (Proposed Standard). Mar. 2012. URL: http://www.ietf.org/rfc/rfc6550.txt.

[Yua+08]  S. Yuan, R. Jedermann, M. Becker, and L. Chen. *Report of WSN Test in Antwerp July 2008*. Tech. rep. SFB637-T4-2008-7. Bremen: Communication Networks, University Bremen, 2008.

[Yua+09]  S. Yuan, M. Becker, R. Jedermann, C. Görg, and W. Lang. "An Experimental Study of Signal Propagation and Network Performance in Monitoring of Food Transportation". In: *Demos and Posters for the Sixth Annual IEEE Communications Society Conference on Sensor, Mesh, and Ad Hoc Communications and Networks. Secon 2009*. Rome, Italy, 2009.

[Ziga]    ZigBee Alliance. *Homepage*. Available at: http://www.zigbee.org. (Retrieved 04/27/2010).

| | |
|---|---|
| [Zigb] | ZigBee Alliance. *Member List*. Available at: http://www.zigbee.org/About/OurMembers.aspx. (Retrieved 07/15/2010). |
| [Zig10] | ZigBee Alliance. *ZigBee Smart Energy 2.0 DRAFT 0.7 Public Application Profile*. June 2010. |
| [Zol13] | Zolertia. *Z1 Datasheet*. Available at: http://zolertia.sourceforge.net/wiki/images/e/e8/Z1_RevC_Datasheet.pdf. 2013. (Retrieved 07/15/2013). |

MIX
Papier aus verantwortungsvollen Quellen
Paper from responsible sources
**FSC® C105338**

If you have any concerns about our products,
you can contact us on
**ProductSafety@springernature.com**

In case Publisher is established outside the EU,
the EU authorized representative is:
**Springer Nature Customer Service Center GmbH
Europaplatz 3, 69115 Heidelberg, Germany**

Printed by Libri Plureos GmbH
in Hamburg, Germany